Shadow Warfare

Security and Professional Intelligence Education Series (SPIES)

Series Editor: Jan Goldman

In this post–September 11, 2001 era there has been rapid growth in the number of professional intelligence training and educational programs across the United States and abroad. Colleges and universities, as well as high schools, are developing programs and courses in homeland security, intelligence analysis, and law enforcement, in support of national security.

The Security and Professional Intelligence Education Series (SPIES) was first designed for individuals studying for careers in intelligence and to help improve the skills of those already in the profession; however, it was also developed to educate the public on how intelligence work is conducted and should be conducted in this important and vital profession.

1. *Communicating with Intelligence: Writing and Briefing in the Intelligence and National Security Communities*, by James S. Major. 2008.
2. *A Spy's Résumé: Confessions of a Maverick Intelligence Professional and Misadventure Capitalist*, by Marc Anthony Viola. 2008.
3. *An Introduction to Intelligence Research and Analysis*, by Jerome Clauser, revised and edited by Jan Goldman. 2008.
4. *Writing Classified and Unclassified Papers for National Security*, by James S. Major. 2009.
5. *Strategic Intelligence: A Handbook for Practitioners, Managers, and Users*, revised edition by Don McDowell. 2009.
6. *Partly Cloudy: Ethics in War, Espionage, Covert Action, and Interrogation*, by David L. Perry. 2009.
7. *Tokyo Rose / An American Patriot: A Dual Biography*, by Frederick P. Close. 2010.
8. *Ethics of Spying: A Reader for the Intelligence Professional*, edited by Jan Goldman. 2006.
9. *Ethics of Spying: A Reader for the Intelligence Professional*, Volume 2, edited by Jan Goldman. 2010.
10. *A Woman's War: The Professional and Personal Journey of the Navy's First African American Female Intelligence Officer*, by Gail Harris. 2010.
11. *Handbook of Scientific Methods of Inquiry for Intelligence Analysis*, by Hank Prunckun. 2010.
12. *Handbook of Warning Intelligence: Assessing the Threat to National Security*, by Cynthia Grabo. 2010.
13. *Keeping U.S. Intelligence Effective: The Need for a Revolution in Intelligence Affairs*, by William J. Lahneman. 2011.
14. *Words of Intelligence: An Intelligence Professional's Lexicon for Domestic and Foreign Threats, Second Edition*, by Jan Goldman. 2011.
15. *Counterintelligence Theory and Practice*, by Hank Prunckun. 2012.
16. *Balancing Liberty and Security: An Ethical Study of U.S. Foreign Intelligence Surveillance, 2001–2009*, by Michelle Louise Atkin. 2013.

17. *The Art of Intelligence: Simulations, Exercises, and Games*, edited by William J. Lahneman and Rubén Arcos. 2014.
18. *Communicating with Intelligence: Writing and Briefing in National Security*, by James S. Major. 2014.
19. *Scientific Methods of Inquiry for Intelligence Analysis, Second Edition*, by Hank Prunckun. 2014.
20. *Quantitative Intelligence Analysis: Applied Analytic Models, Simulations and Games*, by Edward Waltz. 2014.
21. *The Handbook of Warning Intelligence: Assessing the Threat to National Security–The Complete Declassified Edition*, by Cynthia Grabo. 2015.
22. *Intelligence and Information Policy for National Security: Key Terms and Concepts*, by Jan Goldman and Susan Maret. 2016.
23. *Handbook of European Intelligence Cultures*, edited by Bob de Graaff and James M. Nyce—with Chelsea Locke. 2016.
24. *Partly Cloudy: Ethics in War, Espionage, Covert Action, and Interrogation, Second Edition*, by David L. Perry. 2016.
25. *Humanitarian Intelligence: A Practitioner's Guide to Crisis Analysis and Project Design*, by Andrej Zwitter. 2016.
26. *Shattered Illusions: KGB Cold War Espionage in Canada*, by Donald G. Mahar, 2016.
27. *Intelligence Engineering: Operating Beyond the Conventional*, by Adam D. M. Svendsen. 2017.
28. *Reasoning for Intelligence Analysts: A Multidimensional Approach of Traits, Techniques, and Targets*, by Noel Hendrickson. 2018.
29. *Counterintelligence Theory and Practice, Second edition*, by Hank Prunckun. 2019.
30. *Methods of Inquiry for Intelligence Analysis, Third Edition*, by Hank Prunckun. 2019.
31. *The Art of Intelligence: More Simulations, Exercises, and Games*, edited by William J. Lahneman and Rubén Arcos. 2019.
32. *Whistleblowers, Leakers, and Their Networks: From Snowden to Samizdat*, by Jason Ross Arnold. 2019.
33. *Weaponized Marketing*, by Lisa Merriam and Milton Kotler. 2020.
34. *Shadow Warfare: Cyberwar Policy in the United States, Russia, and China*, by Elizabeth Van Wie Davis. 2021.

Shadow Warfare

Cyberwar Policy in the United States, Russia, and China

Elizabeth Van Wie Davis

ROWMAN & LITTLEFIELD
Lanham • Boulder • New York • London

Published by Rowman & Littlefield
An imprint of The Rowman & Littlefield Publishing Group, Inc.
4501 Forbes Boulevard, Suite 200, Lanham, Maryland 20706
www.rowman.com

6 Tinworth Street, London, SE11 5AL, United Kingdom

British Library Cataloguing in Publication Information Available

Library of Congress Cataloging-in-Publication Data Available

Names: Davis, Elizabeth Van Wie, author.
Title: Shadow warfare : cyberwar policy in the United States, Russia, and China / Elizabeth Van Wie Davis.
Description: Lanham : Rowman & Littlefield, [2021] | Series: Security and Professional Intelligence Education Series (SPIES) ; 34 | Includes bibliographical references and index. | Summary: "Examines cyberwarfare policy across three main geopolitical actors and develops an application and deterrence of cyberespionage"—Provided by publisher.
Identifiers: LCCN 2020041366 (print) | LCCN 2020041367 (ebook) | ISBN 9781538149669 (cloth) | ISBN 9781538149676 (paperback) | ISBN 9781538149683 (epub)
Subjects: LCSH: Cyberspace operations (Military science)—United States. | Computer security—Government policy—United States. | Cyberspace operations (Military science)—Russia (Federation) | Computer security—Government policy—Russia (Federation) | Cyberspace operations (Military science)—China. | Computer security—Government policy—China.
Classification: LCC U167.5.C92 D38 2021 (print) | LCC U167.5.C92 (ebook) | DDC 355.4—dc23
LC record available at https://lccn.loc.gov/2020041366
LC ebook record available at https://lccn.loc.gov/2020041367

To Margaret Van Wie
a talented chemist, an amazing scientist on the University of Chicago's Manhattan Project, an officer in the US Navy, and a mother who worked with her generation of women to make the world more accessible for the next generations of women

Contents

Preface

Technology often changes warfare. Cyberwarfare is no exception. Similar to the seismic shift with nuclear warfare, cyberwarfare is modifying warfare into shadow warfare. A few distinctive characteristics of cyberwar emerge as shadow warfare: first, cyberwarfare is continuous and strives to be unseen, making a state of war permanent. This new concept of continuous warfare is addressed by René Girard, who states, "We are thus more at war than ever, at a time when war itself no longer exists." Continuous cyberwarfare also revived the shadow war practices of privateering, nautically used extensively in the fifteenth and sixteenth centuries, a privately owned armed entity permitted by its government to make war on an adversary. Modern privateers interact with entities like principal online firms such as Facebook, Google, Yahoo, and Uber—even the US military—that run continuous security bounty programs, not dissimilar to the Russian and Chinese use of private citizens to cyberattack overseas targets. Privateering allows governments to contract out cyberwar activities, allowing governments to distance themselves even further from the constant conduct of war.

Second, but equally important, cyberwarfare has morphed warfare into shadow warfare by blurring the distinction between adversary and ally. Cyber probes continuously occur between allies and enemies alike, causing cyberespionage to merge with warfare. Espionage, as old as war itself, has technologically merged with acts of cyberwar as states threaten each other with prepositioned malware in each other's cyberespionage-probed infrastructure. These two cyber shifts to warfare are agreed upon and followed by the United States, Russia, and China. What is not agreed upon in this shifting era of warfare are the theories upon which cyberwarfare is based.

This third characteristic of shadow warfare, clashing theories of cyberwarfare, is at least as momentous as the first two. The shift to shadow warfare is tearing at the fabric of war theory, once fairly consistent with conventional armies and mutually assured destruction. Cultures and circumstances develop theories of war. Western militaries, like the United States, follow the legal principles of just war theory and the military philosophy of Carl von Clausewitz. Russian standards of warfare converge at the crossroads of the Russian state under Putin and of the Russian Orthodox Church under a theory of necessity. Chinese politicians and war theorists have brought back a reliance on the traditional theory of Sunzi's *Art of War*, with its basis in Daoism and Confucianism, and a modern Chinese focus on technology. This third characteristic of wide-ranging cyberwarfare theories threatens a rise in violence as the different theories of war make a meeting of the minds difficult.

The United States, the dominant status quo power, reveres the work of Carl von Clausewitz's *On War*. This is especially relevant as the world faces continuous shadow warfare with blurred allies and adversaries. Clausewitz wrote in his first chapter: "War is an act of violence, which in its application knows no bounds." According to René Girard, in *On War and Apocalypse*, Clausewitz identified the unhappy truth that the world tends toward extremes in war. These extremes in war, once thought to have reached the limit first in the total war of the Second World War and then in the threat of nuclear war in the Cold War conflicts of the twentieth century, now ushers in the continuous shadow of cyberwarfare in the twenty-first century. Clausewitz himself took the principles of just war for granted and as not hampering the expanse of war with each new century and each new technological development of war.

Just war theory is now reinterpreted to support cyberwarfare policy. The *jus as bellum* criteria, to consider just a few, include defense to aggression. Cyberattacks are aggressive in preventing a state from meeting basic human needs like access to drinking water or electricity as in currently preset malware. Another principle of just war is the protection of noncombatants. Effective cyberattacks, as with biological warfare, increase the probability of spreading to noncombatants as occurred with Stuxnet. A third principle of just war is proportionality, the idea that it is wrong to cause more harm in defending against an attack than the harm of the attack itself, which was reflected in the US response to Russian meddling in the US election. US cyberwar theory is a progression from the early Western evolution of just war theory and new interpretations of Clausewitz for a new cyber age.

Russian cyberwar theory deviates from the United States, and ultimately from China's, theory first with the Putin government's war theory and second with the "necessary" war doctrine from the Russian Orthodox Church. The Putin government approach to cyberattack relies on earlier Soviet Union theory and practice. This is most evident in Russia's heavy reliance on weaponizing information in disinformation campaigns and using kompromat (meaning compromising material used as black PR) gleaned from cyberespionage and other sources. The long-term president of Russia, Vladimir Putin, starting in his career with the Soviet KGB and then Russian FSB, used kompromat for decades. One purpose of the cyber-released and cyber-obtained kompromat is to undermine governments and officials through powerful and effective disinformation campaigns and recreate Soviet-level status for Russia on the world stage.

The Russian Orthodox Church, often in direct cooperation with the Putin government, also frames Russian cyberwar theory. This reliance on Russian Orthodoxy to create war theory is supported by popular Russian theorist, Aleksander Dugin, who purports to be an Orthodox Christian. Dugin's *Четвертая политическая теория* (*Fourth Political Theory*), has implications for cyberwar theory with his assertion that Russia stands on the side of tradition. Russia and the Russian Orthodox Church, according to Dugin, must reject universalism and insist that the different peoples of the world have to rediscover their own diverse traditions to create a multipolar world. The Russian Orthodox Church, insisting that it is not bound by Western notions of separation of church and state, does not prohibit hostilities if the security of their neighbors and the restoration of trampled justice is at stake. Then war is considered to be necessary, according to the Russian Orthodox Church's Department for External Church Relation's document of the Basis of the Social Contract, VIII. War and Peace. Cyberwarfare theory in Russia is both a reaction against others' cyberwarfare and an active theory to undermine other governments in order to create a multipolar world.

Cyberwar theory in China, like modern China itself, reflects both the new and the old. The new theory is, more of an approach than a philosophy, reflected in China's focus on leapfrogging technologies and its attraction to all things cyber. The old theory is the reliance on Sunzi's *Art of War* (孙子兵法) with some parallels with Russia's refocus on the traditional. China is technologically leapfrogging in an attempt to reemerge as the world player that invented gunpowder and paper. China's big three tech companies—known collectively as BAT (Baidu, Alibaba, Tencent)—and specialized units of the Chinese military are reaching new technological heights as an innovation cyberpower across multiple industries. For instance, China successfully tested the world's first quantum satellite communication, which relies on quantum entanglement physics to exchange provably secure messages.

China's technological leapfrogging is paired with reliance on Sunzi's *Art of War*. Sunzi's classic is especially useful in cyberwarfare, given the book's emphasis on how to fight wars without actually having to do battle. "The art of war is of vital importance to the state." So, begins Sunzi's *Art of War*. Deception is critical in cyber-defense and offense. Sunzi has a lot to say about deception in warfare: "All warfare is based on deception." A central element in Chinese cyberwar theory is to appear to have more when they have less and to appear to have less when they have more. Cyberwarfare theory in China combines the new with the old to reassert China onto the world stage.

With the United States relying on the ever-expanding war conclusions of Clausewitz within just war theory, with Russia combining the Putin government's use of disinformation and the Russian Orthodox Church's injunction that war may be necessary in a quest for a multipolar world, and with China's success in leapfrogging technology and the theoretical assertions in Sunzi's *Art of War*, it is not surprising that there has been an inability to determine a universal set of norms for the policy of cyberwarfare. The clashing theories set the world for an escalation of the twenty-first century's continuous shadow warfare. Once again, the great powers have escalated the terrors of total war and nuclear war with never-ending warfare between each other and the rest of the world.

Acknowledgments

There are always many people to thank in an effort of this sort. First, I would like to thank Anna Pivovarchuk. Not only did she read and comment on every page, but she also added wonderful ideas and suggested areas to expand or clarify. Quite frankly, this book would not be what it is without her. Also, my thanks go to my former graduate assistant Margaret Albert Gullixson, who was with me during the earliest stages of this book. She searched for articles and materials with unflagging perseverance. Her diligence and curiosity made her a pleasure to work with and made her a necessity for thoroughness. She coauthored an early article on shadow warfare with me. Then, I would like to thank the journals who published earlier exploratory efforts on some of the topics discussed in this book, including *Asia Dialogue*, *Fair Observer*, *Asia Times*, and *China Institute: Analysis*. The willingness of these early publishers to support my exploration of these ideas not only made this a better book but also provided the feedback to assure that I was on the right track as I looked at the complexities of shadow warfare and its definitive weapon of cyberwarfare.

Ideas never come out of a vacuum. Oral ideas and written versions all have roots in discussions and interactions with my colleagues. First, I want to thank the talented officers at Marine Base, Camp Smith in Hawaii. I went to give them a lecture on the issues of terrorism in China; while writing my previous book during my almost decade working at the DoD's Asia Pacific Center for Security Studies, they convinced me that my next book—this book—should be on the pressing issues of cyberwarfare. They were absolutely right. Thank you for the enthusiastic recommendation. I am only sorry it took me so long.

Next, when I moved to Colorado School of Mines, I learned that the 1996 Baby Doe cyberattack on a lab at the Colorado School of Mines was the earliest assault cited in US military documents in a two-year-long Russian cyberespionage operation, dubbed "Moonlight Maze." The cyber operation entered numerous university research facilities and defense contractors, unclassified computer networks at the Energy Department's nuclear weapons and research labs, and at NASA. Moonlight Maze systematically compromised the Pentagon's main unclassified computer system, the Non-Classified Internet Protocol Router Network (NIPRNET). In response, the US Pentagon ordered new encryption technology, intrusion detection devices, and computer firewalls. The Russian origin of the cyber operation was confirmed at a Senate subcommittee in October 1999 in the first public confirmation of Moonlight Maze. So, my institution, the Colorado School of Mines, was among high-priority targets of early cyberattacks and an inspiration for this book.

Consequently, I would like to thank my colleagues at the Asia Pacific Center for Security Studies, where the book began its journey, and my colleagues at the Colorado School of Mines, where the book reached its journey's end. All of my colleagues have been important in shaping my ideas, but special thanks go out to Joan Johnson-Freese, Robert Wirsing, Virginia Watson, Ehsan Ahrari, Kathleen Hancock, Derrick Hudson, James Jesudason, and Zhu Qin. They have all made me a better scholar by sharing their ideas and challenging mine. They pushed my thinking into new areas and forced me to consider new interpretations. They discussed and debated these topics with me for hours. I am deeply grateful. I am also appreciative of the friendship and wisdom of my colleague Dinesh Mehta, who is a brilliant computer science scholar and professor. Deserving extraordinary thanks is my colleague and friend Dr. Carl Mitcham, who generously offered his time, his insight, and his kindness in reading an early draft and pushing my analysis to deeper levels. All of the mistakes and errors are my own.

Finally, and most importantly, my enthusiastic thanks go to my family: my wonderful spouse and partner, Greg Davis, who has provided support in so many ways but especially by taking care of me; to Kate Davis, my talented and wonderful daughter, who inspires me to be a better academic and scholar by her own example; her spouse Dmitry Golomidov, a talented computer scientist, who was willing to discuss Russia and computers with me; and to my son, James Davis, who gave me early primers and discussed this topic with me for years. And to my mother, Margaret Van Wie, to whom this book is dedicated, words cannot express my gratitude for all you gave to me, our world, and succeeding generations. To all of you, I give my love and thanks.

Chapter 1

Shadow Warfare

In 2013, a small dam in the state of New York found itself under cyberattack. Iranian cyberattackers infiltrated the offices of the site's personnel on Blind Brook in Rye Brook, New York, but did not get as far as the dam's controls. However, the realization that a facility 30 miles away from New York City could flood its infrastructure, causing untold havoc in one of the world's biggest hubs, shocked the American public and drew officials' attention.

Cyberwarfare is a major weapon, joining drones, satellites, and robots in the high-tech revolution that has made shadow warfare so prevalent in the twenty-first century.[1] Shadow warfare is intentionally difficult to see because it hides in modern everyday technology like the internet and computers, creating a permanent low-intensity state of war. Within the era of shadow warfare, the threat landscape has altered seismically in the past three decades, and the cyberattacks reported in the media are just the tip of the iceberg. Cyberwar, nonetheless, is king.

This new era of warfare, like the other eras of warfare before it, is shaped by the weapons that drive it. A few distinctive characteristics of cyberwar emerge as shaping shadow warfare. First, cyberwarfare is continuous and strives to be unseen or untraceable, making a state of war permanent. So, while the footprint[2] of any one attack may be smaller, warfare is now an ongoing phenomenon. This new concept of continuous shadow warfare is addressed by René Girard, who states that "we are thus more at war than ever, at a time when war itself no longer exists."[3]

Continuous shadow warfare also revived the practices of privateering, used extensively in the fifteenth and sixteenth centuries on the high seas, now meaning unacknowledged private cyberattackers. Principal online giants like Facebook, Google, Yahoo, and Uber—and even the US military—run continuous security bounty programs not dissimilar to the Russian and Chinese use of private citizens to launch cyberattacks on overseas targets in an attempt to create plausible deniability. It is permanent low-intensity warfare that the principal actors often deny entirely.

Second, but of great importance, is the fact that cyberwarfare has morphed warfare into shadow warfare by blurring the distinction between adversary and ally. A chief method by which allies and adversaries merge is through cyberespionage. Espionage, as old as war itself, has technologically integrated with acts of cyberwar as states threaten each other with previously positioned malware in each other's cyberespionage-probed infrastructure. Moreover, as cyberespionage merges with warfare, cyber probes continuously occur between allies and adversaries alike. Given the relatively low cost of cyberwarfare, permitting both great powers and lesser powers into its arena, the temptation to control allies has become almost as strong as the need to understand and manipulate adversaries.

Cyberwarfare has evolved from the synthesis of new technologies and military imaginations—and the opaque domestic laws governing them—so that not only is declared war a thing of the past, but also traditional concepts of adversaries and allies fade.[4] These two cyber shifts to warfare—permanent low-intensity conflict and the merging of both warfare with espionage and of allies with adversaries—are agreed upon and followed by the United States, Russia, and China. What is not agreed upon are the theories on which cyberwarfare, as the primary instrument of shadow warfare, is based.

This third characteristic involving different theories of cyberwarfare is at least as momentous as the first two. The shift to shadow warfare is tearing at the fabric of war theory, once fairly consistent with conventional armies and mutually assured destruction. Cultures develop theories of war, and circumstances mutate them. The new technological circumstance of cyberwarfare is causing states to either adapt existing theory to new circumstances or avoid constraints on military behavior altogether. Western militaries, like the United States, use war theory that follows strict legal principles to conform to public scrutiny and rely upon military lawyers, as well as following the military philosophy of Carl von Clausewitz.[5] Russian theories of warfare meet at the convergence of the state under President Vladimir Putin and the Russian Orthodox Church, which uses a theory that war is never just but is sometimes necessary. And Chinese politicians and war theorists have brought back a reliance on the war theory of Sunzi's *The Art of War*, with its traditional basis in Daoism and Confucianism, and its new focus on technology. This third characteristic of wide-ranging cyberwarfare theories threatens a rise in violence unseen since the Cold War as the different theories of war meet in the uncertain circumstances and technologies of shadow warfare.

The shift to shadow warfare in the twenty-first century has parallels with the shift to nuclear warfare in the mid-twentieth century, as stated by Richard

Clarke, former US national coordinator for security, infrastructure protection, and counterterrorism. Clarke writes: "The US developed and systematically deployed [new types of weapons], based on our new technologies, and we did so without a thoughtful strategy. We created a new military command to conduct a new kind of high-tech war, without public debate, media discussion, serious congressional oversight, academic analysis, or international dialogue."[6] As many developments in shadow warfare are moving within shifting and uncertain strategies and theories, there is no push to seriously constrain the new technology that has enabled it. The reluctance to constrain the new war technology reflects a world increasingly dependent on enhanced communication abilities presented by cyberspace and on the new way that technology can enrich quality of life.[7] In this environment, shadow warfare has thrived and caused instabilities,[8] not only for major powers like the United States, Russia, and China but also around the world's major industries like banking and energy, as well as in adversarial governments like Iran by disrupting nuclear programs or by disrupting elections in allied governments like Germany.

The move toward shadow warfare began in the 1990s as more governments, militaries, universities, industries, commercial bodies, banks, and private citizens became increasingly connected to, and dependent on, the internet. The next decade brought a spike in reported cyberattacks and cyberespionage with sophisticated weapons like Stuxnet, Flame, Duqu, and Gauss—and these weapons are merely the ones exposed in open-source materials. Cyberespionage also began targeting defense industries as well as traditional government-to-government cyberespionage operations. While cyberespionage is routinely denied, deniability is an important aspect of shadow warfare. The United States, China, and Russia—three of the major players in shadow warfare utilizing cyberweapons—not only develop both defensive and offensive capabilities in cyberspace[9] but are also using cyberattacks and cyberespionage against each other, targeting adversaries and allies alike.[10]

When cyberattacks occur concurrently with traditional military movements like tanks rolling across borders, the union is classified as "hybrid" warfare. In these cases, cyber conflict joins with high-intensity or kinetic conflict to achieve multipronged warfare. A classic case of hybrid warfare happened on August 8, 2008, in Georgia's breakaway region of South Ossetia, between Georgia and Russia. Other hybrid attacks are part of acknowledged warfare: Israel's 2006 war against Hezbollah, in which Tel Aviv alleged that cyberwarfare was part of the conflict; Israel's airstrike (in September 2007) on the largest city in eastern Syria, Dair Alzour, the site of the suspected Al Kibar reactor, in order to destroy a budding nuclear capability, coordinated with a cyberattack designed to interfere with the computers of the Syrian integrated air defense systems; and so forth.[11] For cyberwarfare to work as a part of active military operations, its targets have to have accessible vulnerabilities that can be exploited in ways the attacker finds useful.[12]

Unlike cyberwarfare that accompanies recognized warfare, there are instances where cyberattacks are part of persistent low-grade shadow warfare between undeclared adversaries. Classic examples of cyberattacks taking place not between traditional adversaries include the April 2007 cyberattack on Estonian ministries, banks, and media in the wake of relocation from the town square to a military cemetery of the Bronze Soldier of Tallinn—a symbol of Soviet repression for ethnic Estonians and a symbol of Soviet triumph over the Nazis for ethnic Russians—with the largest part of the attacks coming from Russia. In another instance, the website of the Kyrgyz Central Election Commission was defaced during its 2007 election. In March 2009, a cyberespionage network dubbed GhostNet, using servers mainly based in China, tapped into classified documents from government and private organizations in 103 countries, including the computers of Tibetan exiles. Most notoriously, in September 2010, Iran's Natanz nuclear enrichment facility was attacked by the Stuxnet worm in an attempt by the United States and Israel to slow down the Iranian nuclear centrifuges from refining uranium into a usable form. None of these cyberattacks either accompanied kinetic attacks or occurred between acknowledged adversaries; all of these sought to be anonymous.

To people on the ground, the reality of these shadow conflicts is impactful, not least because these attacks are happening in places where war has not been, and will not be, declared.[13] Nowhere has the issue played out more directly than in Miram Shah, in northwestern Pakistan. It has become a fearful and paranoid town after shadow warfare dealt multiple drone strikes and the collapse of most electronic communication systems. Communication—which families need in order to reach relatives sending home remittances from abroad—was made almost impossible because cell phone networks were disabled by an unseen electronic attacker, while internet cafes were shut by a known drone attack.[14] Two of the major players in this sometimes invisible conflict were US attackers and Pakistani targets—supposed allies—as the US war in Afghanistan spilled over into northwestern Pakistan's provinces where al-Qaeda fighters were hiding.

A PERMANENT STATE OF WAR

One of the defining forms of shadow warfare is cyberattacks—attacks that occur in cyberspace. Cyberspace—both physical and virtual—is subject to various definitions depending on the needs and the perspective of the person who defines it.[15] Some definitions emphasize the use of cyberspace, such as the one that has broadly evolved in national security and international relations. This category of definitions emphasizes that cyberspace is a domain—albeit a manmade one—similar to land, sea, air, and space.[16] Other definitions emphasize the technical infrastructure of cyberspace, such as that used by the US Department of Defense: "A global domain within the information environment consisting of the interdependent network of information technology infrastructures, including the internet, telecommunications networks, computer systems, and embedded processors and controllers."[17]

Expanding from the definition of cyberspace are cyber conflicts. As defined in the initial Cyber Conflict Studies Association research agenda in 2005, cyber conflict "is the conduct of large scale, politically motivated conflict based on the use of offensive and defensive capabilities to disrupt digital systems, networks, and infrastructures, including the use of cyber-based weapons or tools . . . for political ends"[18] conducted by both state and non-state actors against a variety of targets. Under this broad umbrella term of cyber conflict is a wide variety of cyberattacks, some of which are weapons of war—often shadow warfare weapons—and others are cyberattacks of various types.

Cyberwar

Cyberwarfare[19] is about much more than the internet and involves commercial networks and discreet national networks.[20] The motivations to engage in shadow warfare are as they always have been in war: the pursuit of power and perceived national interest. Warfare is the conduct of war, but it does not exclusively use military means.[21] In shadow warfare, the operational and tactical battlefield attempts to gain power of information and influence through the use of cyberweapons and cyber assets. However, it is hard to duplicate traditional notions of victory and defeat in cyberspace because the cyber domain is more complex and populated with a more diverse set of actors, generating a chaotic sphere of conflict. A common framework of policies, norms of behavior, and values—including the use of precise and proportionate force in the cyber domain while avoiding unintended consequences—is a challenge.

The relative newness of cyberwarfare means it is still emerging as a doctrine. This was similarly true of air power and nuclear deterrence when they were being developed decades ago and before they became enshrined in doctrine. So, prudence suggests that similar deliberations should be anchored in the accepted wisdom of war by old masters. What counts as warfare? Chinese philosopher-general Sunzi says: "The art of war is of vital importance to the state. It is a matter of life and death. . . . The art of war, then, is governed by five constant factors . . . The Moral Law; Heaven; Earth; The Commander; Method and Discipline."[22] Prussian military theorist Carl von Clausewitz still offers the most concise answer. First, all acts of war are violent or potentially violent. Second, acts of war are a means to compel the adversary to accept the attacker's will. And, finally, to qualify as an act of war, an attack must have some kind of political goal or intention.[23] As cyberwarfare is a new form of shadow warfare, not all preexisting terms and concepts fit neatly, but the general concepts prevail.[24]

Shadow warfare uses cyberwarfare for reconnaissance, information operations, the disruption of critical networks and services and to complement electronic warfare and information operations. The original approach to cybersecurity from the 1990s in the United States meant setting up a national Computer Emergency Response Team (CERT), assigning responsibility to science ministries and creating specialized units within the national police. Some states included specific plans for informational and political operations. Others linked cyberwarfare capabilities with existing electronic warfare planning. The linkages between electronic warfare and cyberwarfare have expanded as computer networks become increasingly mobile and wireless.[25]

The doctrine and structure of cyberwarfare are rapidly changing as a result. For example, states now consider their digital infrastructure—electric and power grids, online banking and services, records, and communications—to be strategic national assets.[26] Several states have consolidated their domestic offices that are involved in cyberwarfare as well as increased communication between these offices. For example, the US Cyber Command was formed in May 2010 after a significant breach of Department of Defense (DoD) networks in November 2008 at US Central Command, and South Korea's military formed its Cyber Command in early 2010 following several cyberattacks by North Korea (DPRK).[27] Moreover, states are assessing critical infrastructure as needing protection, including industrial defense bases, financial systems, transportation networks, electrical systems, nuclear power plants, and water works.

Cyber Defense

Cyber defense[28] essentially consists of two basic strategies. The first strategy, deterrence through denial, is the demonstrated capability to achieve the defense of a stated interest. The second strategy, deterrence through punishment, is to inflict such a high cost on the attacker that the attack is not worth the effort even if it achieves its goal.[29] Deterrence through denial is primarily a defensive contest. Deterrence through punishment is primarily an offensive contest, based on the threat of credible and painful retaliation for attacks.

The relationship between cyber defense through punishment and escalation control is complex and ultimately unsatisfying. On the one hand, credibility is perversely bolstered when the adversary fears that retaliation will be total and relentlessly destructive. On the other hand, the adversary needs to believe that, should they seek to sue for peace, the retaliator has the ability to de-escalate its attacks and even terminate them.[30] Although only a handful of countries have the capability to carry out high-caliber cyberattacks, over 100 countries have begun to organize their national cyberwarfare units for defensive purposes.

Cyber Offense

A cyber offense refers to deliberate actions to alter, disrupt, deceive, degrade, or destroy computer systems, cyber information, or programs,[31]or to use cyber means to attack, disrupt, or destroy physical objects by use of a cyber delivery system. Cyber offense capabilities can be dramatically cheaper than effective cyber defense capabilities since the cyberattacker only needs to find one way in, but the cyber defender needs to stop every possible avenue of entry. Some sophisticated cyber offense weapons that have been uncovered, like Stuxnet, are elaborate operations that cost considerable amounts in terms of hours and planning.

Cyberwarfare carries complications. For instance, states sometimes work through third parties to decrease risk and increase plausible deniability.[32] Another complicating factor is cyber counteroffensives—for example, between the United States, China, and Russia, which all have deep relationships with each other on other strategic fronts—can cost[33] unforeseen consequences of military retaliation or to bilateral trade.[34] Another serious consequence of cyberwarfare is the difficulty of containing cyber malware—keeping malware from escaping into the "wild"—similar to the inability to contain chemical warfare or nuclear fallout. In a quest to create a smaller, near-invisible footprint, the consequences of cyberwarfare should be factored into decisions when creating cyber defense and cyber offense strategies.

As shadow warfare develops and matures, cyberweapons become increasingly effective and sophisticated. Early DDoD—distributed denial of service—cyberattacks are still used but are increasingly replaced by more specific cyberweapons, like using phishing techniques to steal data or sending malware, using social media to spread disinformation, or gaining control of external operating systems. To fully exploit the benefits of cyberweapons, they remain secretive—to gain maximum value with added deniability, much like Israel never admitted to either the cyberattack on, or the kinetic bombing of, Syria's nuclear development site.

The problem with shadow warfare in liberal democracies is that democratic societies not only no longer officially declare war, but now the secretive and permanent nature of cyberwarfare means that there is no democratic discussion on the value of conflict to society. For instance, the United States was not at war when it launched an attack on communication systems in a Pakistani village despite the fact that Pakistan was an ally—and, more importantly, without the knowledge of Congress, the wider public, or open debate and discussion on the merits of the operation. In another instance, the United States had burned its bridges with the now confirmed false pretext of weapons of mass destruction in its 2003 invasion of Iraq when it decided to use a secretive cyberattack in Iran. US citizens were not informed of another venture in the Middle East until years later.

Stuxnet

The most sophisticated cyberweapon unmasked has been Stuxnet. Stuxnet is the second phase of the Operation Olympic Games that targeted Iran's nuclear infrastructure and created a specific cyberweapon that was Stuxnet. It was a cyberattack on a type of industrial controller at Iran's nuclear enrichment plant in Natanz, causing 1,000 centrifuges to spin out of control.[35] This was the first major cyberweapon that caused physical destruction rather than destruction to another cyber system or theft of electronic data.[36]

As David Sanger describes in *The New York Times*, the concept of the code-named Olympic Games and the cyberweapon Stuxnet was initiated under the administration of George W. Bush. With a limited number of good options to deal with Iran's resumption of enriching uranium—Europe was divided on what imposing sanctions on Iran could cost its economy, and the United States had little credibility after first falsely accusing Saddam Hussein of reconstituting his nuclear program and then resorting to traditional warfare in Iraq—secretly creating a cyberweapon seemed like a credible option to stop Iranian nuclear ambitions.[37]

The goal of this cyberweapon was to command the Natanz plant's industrial computer controls. Access required leaping the air gap—which physically separates the facility's computer system from the wider connectivity of the internet—that isolated Natanz. The leap was first conducted via a specially infected USB drive and later through more sophisticated methods. To gain control of the computers commanding the giant centrifuges that spin at tremendous speeds in order to enrich the fissile isotope, Uranium-235, which can be used in both nuclear power plants and nuclear weapons, an initial cyberespionage tool was employed to outline the centrifuge network, describing the structure and workings of the enrichment plant, complete with maps of the electronic directories of the controllers and how they were connected to the centrifuges deep underground.[38]

The cyberweapon was created by the National Security Agency (NSA) with the premier Israeli cyber team, Unit 8200. The United States and Israel collaborated on this cyberattack for at least two reasons. First, Israel's military Unit 8200 has a renowned cyber capability as well as considerable intelligence about operations at Natanz that would be vital to making the cyberattack a success. Second, the United States wanted to dissuade the Israelis from carrying out their own preemptive strike against the Iranian nuclear facilities and thus needed to convince Israel that the cyberattack would be effective by involving them in the program.[39]

The cyberattack was launched in 2008, with the centrifuges spinning out of control. Initially, the Iranians were confused about the cause. This was partly because no two attacks were exactly alike, and partly because the virus sent signals to the Natanz control room indicating that everything in the centrifuge area was operating normally. By the time the Iranian side realized the attack was taking place, the cyberweapon had successfully wrecked centrifuges and disrupted the enrichment of Uranium-235.[40]

Sanger outlines how Stuxnet transitioned from the Bush administration to the Obama White House. Although President Barack Obama came to office with an interest in cyber issues, he learned the art of cyberwar while in office. The Obama administration took over the Olympic Games operations and continued the attacks on Iran's nuclear program.[41]

Then, in the summer of 2010, the Stuxnet malware escaped from the Natanz uranium processing facility. The malware spread to an engineer's computer when it was hooked up to the centrifuges. When the engineer left Natanz and connected his computer to the internet, the malware virus began replicating around the world.[42] Even with the Stuxnet virus replicating itself "in the wild," computer security experts began dissecting it and determining its purpose as the cyberattacks under the Olympic Games program continued. Another version of the computer virus brought down just under 1,000 centrifuges. Operation Olympic Games was still on.[43]

NOT ALL CYBERATTACKS ARE CYBERWARFARE

Attacks in cyberspace are not only government-based but can be private as well. Private attacks only sometimes garner a government response and are usually dealt with between private entities. There are at least four major private groups that engage in cyberattacks: criminals looking to profit from e-commerce, corporations that spy on each other to gather trade secrets and technology, social groups/hackers that have specific agendas—called hacktivism—and more malicious groups that engage in some level of cyberterrorism. These private entities might be easier to track and capture compared to government-sponsored groups of attackers primarily because they do not have government-size coffers to spend on cyberattacks.

In addition to the economic motivations of corporations, some companies are now taking a more sophisticated strategy when it comes to cyberattacks, admitting that the criminals are going to get into their systems, reengineering their defenses to protect the vital data within their networks, and trying to catch the criminals once they are inside the networks.[44] These private attacks are too often confused with, or actually linked to, cybercrime, cyberterrorism, and cyberespionage.[45] Cybercrime, with a primarily political focus, is a good place to start.

Cybercrime

Cybercrime can impose economic costs far out of proportion to the price of launching the attack.[46] While the 2017 annual cost of natural disasters in the United States is estimated to be $306 billion,[47] cybercrime damage costs are estimated by industry giants like Dell, Verizon, and Malware Bytes to be in the trillions.[48] This represents the greatest transfer of economic wealth in history, risks losing the incentives for innovation and investment, and will be more profitable than the global trade of all major illegal drugs combined. In response to escalating cybercrime, cybersecurity spending is expected to exceed $1 trillion from 2017 to 2021.[49]

Cybercrime can merge into government-sponsored cyberattacks. For instance, writers in some of China's military journals speculated that cyberattacks could disable American financial markets. The dilemma for this kind of attack is that China is as dependent on the same financial markets as the United States and could suffer serious domestic economic consequences for inflicting any large-scale attack of this sort. Additionally, most major financial institutions have back-up systems that have them back online in a matter of days. However, sometimes this type of cybercrime may be worth the costs, as was the case in Estonia in 2007, which was likely very inexpensive for Russia—especially compared to the costs for Estonia, which included the shutting down of banks and ATMs, media outlets, and most government functions. Moreover, such cyberattacks could potentially be a useful tool for private groups, like the ransomware created for monetary gain. Examples of this include the attack on Travelex, a company that provides banks such as Barclays with cash, or governments that reject the global market economy, like North Korea's cyberattack capability operating in a physical world in which North Korea barely participates.[50]

Some famous cybercrimes have been conducted by the government of North Korea. The DPRK's most spectacular cyberattack took place in 2014 against Sony Pictures Entertainment to block the release of a political farce movie, *The Interview*, which satirized an attempt to "kill" North Korea's leader, Kim Jong-un. What has been less publicized is that the DPRK also unconventionally attacked a British television network a few weeks earlier to stop the broadcast of a drama about a nuclear scientist kidnapped in Pyongyang. This type of unconventional cyberattack is different than most countries' cyber strategy but similar to cyberattacks on a South Korean television station in 2013 in retaliation for negative coverage of the North Korean leadership.[51]

The DPRK has also conducted a series of cybercrimes to both disrupt the international system and gain much needed foreign currency. US intelligence officials linked North Korea to the WannaCry ransomware attack in May 2017. The attack infected more than 230,000 computers in over 150 countries for a few weeks and involved an outbreak of malware that encrypted files which were offered to be released in exchange for Bitcoin payments.[52]

Hacktivism

Cyber methods used to promote political or social aims and to destroy political opposition—hacktivism—is another form of private cyberattacks. Many targets of hacktivism are of an overtly political nature. Lulz Security, commonly referred to as LulzSec, a well-known hacktivist group, once successfully damaged the websites of the US Air Force, the US Senate, the CIA, and the UK's National Health Service (NHS), along with many others, before its members began to be arrested and sentenced. LulzSec accessed the US Air Force website, released secure information from the Senate, shut down the CIA site for a few hours, and attacked NHS online services.[53] The alleged purpose of the hacktivist group was purportedly to embarrass prominent entities for their lack of cybersecurity.

Other countries, like Portugal and Egypt, have also suffered from unauthorized access to their online systems. In Portugal, in response to the brutal suppression of public protests held on November 24, 2011, against austerity measures, the websites of the Bank of Portugal, the Portuguese parliament, and the Ministry of Economy, Innovation, and Development were all attacked.[54] The Egyptian government shut down the internet for a few weeks in February 2011, at the start of the Arab Spring. In response, Google and Twitter created a Speak2Tweet telephone service, active from 2011 until 2015, that allowed anyone to leave an audio message that then showed up on Twitter while the user remained anonymous. The Egyptian government then shut down cell phone networks.[55]

Other examples of hacktivism against states include Iran and Turkey. For instance, in 2009, Iranians protested unsuccessfully against perceived widespread election fraud, inspiring Anonymous—the most prominent hacktivist group that opposes internet censorship and supports vigilantism on many international issues—to set up an information exchange website called Anonymous Iran. An extremist hacktivist group, RedHack, uses highly organized cyberattacks and leaks of information online to criticize the Turkish government's swing to authoritarianism. Previous targets of RedHack include the Turkish Council of Higher Education, the state's police force, the army, Türk Telekom, and the National Intelligence Organization.[56]

Another example of hacktivism concerns the activities of hacktivist group Anonymous Africa. During the 2013 reelection of Zimbabwe's president, Robert Mugabe—a vote that was criticized as rigged[57]—Anonymous Africa used hacktivism to shutter the government newspaper and fifty websites, including those associated with the ruling party, insisting that Mr. Mugabe's regime had extensive airtime on state TV to support its message while giving none to the opposition. Subsequently, the website of South Africa's Independent Media—the largest group of newspapers in the country—underwent a DDoS attack following the publication of a pro-Mugabe opinion piece.[58]

A major issue with hacktivism is whether the perpetrator is public—a state—or private—an individual or group of individuals. One example is the Shamoon hacktivist attack. Saudi Arabia's national oil company, Saudi Aramco, confirmed reports that its computer networks were shut down in August 2012 by a hacktivist attack. The computer security firm Symantec announced that the malware made any infected computer unusable by wiping clean sectors of the hard drive. The hacktivist attack weapon was dubbed Shamoon.[59] It is not clear where the Shamoon cyberattack originated. An unknown hacktivist group, alternately calling itself the Arab Youth Group or the Cutting Sword of Justice, which could be either two different groups or just two different names, claimed unverified responsibility for the attack. The hacktivists decried Saudi leaders for their ties to the United States and for working with the Israelis in trying to thwart Iran's nuclear ambitions.[60]

This successful breach was followed by two similar successful unauthorized accesses at the end of 2016 using malware dubbed Shamoon 2. The attacks may have been in retaliation for cyberattacks on Iran that forced it to cut internet conductivity with its oil industry. US intelligence officials and cybersecurity firms suggested that the cyberattacks were conducted by the government of Iran, although no specific evidence was offered to support that claim. If this was indeed a state attack, it is more honest to label it a state-sponsored cyberwar rather than hacktivism, which presumes a private individual or group of individuals, not individuals hired by a state to do its dirty work.[61]

Some states do use privateers or criminals to conduct their attacks, but cyberattack by proxy is still a cyberattack. Ties between governments and privateers or criminal proxies are some of the darker aspects of cyber operations. For instance, while Russia occasionally uses privateers to maintain a degree of separation from its cyber operations, some supposed privateers are actually Russian military and civilian operatives, including CyberBerkut, the Cyber Caliphate, and Guccifer 2.0, as illustrated in chapter 2. Other instances seem to be privateers doing piece work as needed. The relationship between the privateers and the government can be primarily financial. In this instance, Russian intelligence officers would pay a privateer approximately $100 for each compromised email account.[62] Privateering is still used by states, but it is becoming a less effective manner of shielding states from responsibility for conducting cyberattacks as governments enhance their ability to identify the perpetrators of cyberattacks and cyberespionage.[63]

Cyberterrorism

Cyberterrorism is often discussed as a new factor in shadow warfare because cyberattacks can have a low cost that makes it available to non-state entities. Cyberterrorism is the use of cyberattacks in order to create terror incidents. In reality, cyberterrorism occurs very rarely. Cyberterrorism activities are a bridge between private and state-based cyberattacks. This is because cyberterrorism can be carried out between private entities, or between private entities and state actors, or between governments. Terrorists and terrorist organizations do make and move money over the internet, but this may well be cybercrime rather than cyberterrorism.[64] It is a commonly accepted principle that despite the heinous nature of their acts, terrorists—cyber or otherwise—should be prosecuted under the same criminal law as any other transgressor. The defense of lawful processes is a core value of the United Nations and a fundamental pillar of the rule-of-law approach to the fight against terrorism.

On a private level, cyberterrorism is to date less effective and less disruptive than physical terror attacks. The advantages of cyberterrorism are that it is cheaper and easier to carry out than a physical attack and can be done from remote locations outside of the state under attack. The disadvantages of cyberterrorism are the lack of the dramatic loss of life and the visible threat that causes the desired terror. Cyberterrorists do use the internet to steal credit card numbers or valuable data to provide financial support for their operations. Because of this, cyberterrorism has attracted considerable attention, but so far cyberterrorism has not resulted in direct terror activities. Rather, cyberspace has offered a safe haven[65] for propaganda, intelligence collection, financing operations, or hacktivism.[66]

On a state level, cyberterrorism can attempt to cause social devastation through physical destruction—perhaps by causing dams to overflow into populated areas or attempting to cause explosions in nuclear power plants in order to turn them into dirty bombs. Cyberterrorism can attempt to cause social disruption—perhaps by shutting down critical national infrastructure grids like energy, transportation, or banking. The purpose of such cyberterrorism might be to coerce and intimidate a government or to cause panic and fear among citizens.[67] State-based cyberterrorism generally warrants a state response and can vary in scope from espionage to hybrid or physical warfare. From a strategic military perspective, if a cyberattack does not cause damage that rises above the threshold of routine disruptions, it need not pose an immediate or significant risk to national security.[68] James A. Lewis of the Center for Strategic and International Studies wrote an enlightening comparison of physical and cyberattacks on hydroelectric dams in which he concluded that since cyberattacks may not destroy the actual infrastructure, it might be less damaging than a physical attack.[69] While this does not preclude that cyberterrorism can cause real harm, so far it is a smaller threat to states than cybercrime, hacktivism, or state-sponsored cyberattacks.

MERGING WARFARE WITH ESPIONAGE AND ALLIES WITH ADVERSARIES

Cyberespionage is increasingly merging with cyberattacks and cyberwarfare. Cyberespionage within democracies treads a thin line between providing security and eroding public trust as demonstrated by the controversy over the NSA's collection practices inside the United States.[70] Cyberespionage between states has dramatically increased espionage in the era of shadow warfare, raising the issue of cyberespionage as the first level of attack before cyberweapons are employed to new levels. For instance, spear phishing—an email scam intended to steal data or to install malware—can be used for cyberespionage or cyberattacks.

Espionage is an ancient art, one that is instrumental to the formulation of strategy and tactics that can only be based on reliable intelligence about allies and adversaries. The issue with espionage is not the potential advantage that is to be gained. The issue with espionage is whether it is consistent with the laws of war and the policies that underlie these laws. The preliminary effort appeared to maintain a balance of espionage and attack that was proportionate to the problem intended to be addressed, but the lack of transparency in cyberattacks or cyberespionage means that it is extremely difficult to confirm if the rule of proportionality is routinely followed.

The United States and others want to differentiate between the universal practice of foreign intelligence gathering through cyberespionage and commercial espionage through cyberespionage. This view regards commercial espionage as illegitimate. China and others suggest that this is an exaggerated distinction to deflect attention away from the reality that most states do both.[71] For instance, the United States complains that China and Russia are using cyberespionage to steal trade and technology secrets.[72] Given the massive amount of information and research on computer networks, cyberespionage can collect data quickly and with little risk, according to a report by the US National Counterintelligence Executive titled "Foreign Spies Stealing US Economic Secrets in Cyberspace."[73] The report alleges that intelligence services, private companies, academic institutions, and citizens of dozens of countries target the United States, although the report only openly names China and Russia.[74] The allegation is that by using cyberespionage to boost the attacker's economies—and by extension harm the host economy—the theft poses a threat to national prosperity and security. Clearly, finding and reading restricted information is intrinsically interlinked to using that information for one's own national benefit. Cyberespionage and cyberattacks can be hard to untangle.

Cyberespionage is not limited to adversaries. The scandals of 2013 and 2015 in Germany demonstrate the role of cyberespionage even among close allies. Although German chancellor Angela Merkel stated that "spying among friends, that simply isn't done,"[75] it is, in fact, done all the time. In 2013, it was revealed that the NSA was spying on Germans, including an unproven claim that Merkel's cell phone was being monitored. Partly as a result of inquiries into NSA cyberespionage, in 2015, the German press uncovered that the country's foreign intelligence service, the BND—and not the BfV, which is the domestic intelligence service—was monitoring German allies through the use of computer search terms aimed at European political leaders and businesses. [76]

In a similar incident, European officials were angry that the US agencies had monitored the offices of the European Union in New York and Washington, based on information in documents obtained by Snowden.[77] In March 2017, in perhaps the largest leak of CIA documents in history, WikiLeaks released thousands of pages describing sophisticated cyberespionage tools and techniques used to break into smartphones, computers, and even internet-connected televisions. The documents include instructions for compromising common computer tools for use in spying, including the online calling service Skype, Wi-Fi networks, documents in PDF format, commercial antivirus programs used to protect personal computers, and ways to steal passwords using the autocomplete function on Internet Explorer.[78] Complicated by the convoluted and constantly evolving nature of cyberespionage, the sometimes contradictory goals between freedom and security need to find, and maintain, a balance.[79]

In any case, cyberespionage is increasingly hard to distinguish from cyberattacks. After the Olympic Games used cyberespionage to gather intelligence and lay markers—before the formal cyberattack via Stuxnet and other Stuxnet-like malware to disrupt and delay Iran's nuclear program—it was clear that cyberespionage was intrinsically linked to the deployment of cyberweapons. Some of these operations were subsequently discovered, namely

Flame, Duqu, and Gauss, although it is almost certain that there are multiple other cyberweapons, cyberespionage, and cyberattacks occurring on a daily basis around the world.

Flame

Flame is a sophisticated computer virus used for cyberespionage by mapping networks and collecting data, including keystrokes, audio, and visual snapshots developed as part of Olympic Games. The Flame virus—occasionally dubbed Skywiper or Flamer—is a large and complex piece of malware for cyberespionage.[80] The NSA and CIA, together with Israel's military, jointly developed the massive computer virus that secretly mapped and monitored Iran's computer networks. Details about Flame provide clues as to what may be the first sustained campaign of cyberespionage and cyberwarfare against an adversary of the United States.[81]

The Flame virus was designed to replicate across even highly secure networks, then control regular computer functions to send information back to its creators. The code could activate computer microphones and cameras, log keyboard strokes, take screenshots, extract geolocation data from images, and send and receive commands and data through wireless technology. Flame was exceptionally large, with 20 megabytes of code. For scale, a 70,000-word book without pictures or formatting is about 0.4 megabytes.[82] Though malware's size is not an exact measure of sophistication, in this case size suggests that it took a lot of time and work to create.

Flame uses five encryption methods, three compression techniques, and at least five file formats. It was designed to do all this while masquerading as a routine Microsoft software update and evaded detection for years by using a sophisticated program to crack the encryption algorithm. "This is not something that most security researchers have the skills or resources to do," said Tom Parker, chief technology officer for FusionX, a security firm that specializes in simulating state-sponsored cyberattacks. "You'd expect that of only the most advanced crypto mathematicians, such as those working at NSA."[83] According to cryptographic experts, Flame is the first malicious program to use an obscure cryptographic technique, known as prefix collision attack, which allowed the virus to fake digital credentials that had helped it to spread.

Flame shows the importance of mapping networks and collecting intelligence on targets as a prelude to an attack, especially in closed computer networks. Gaining and keeping access to a network is the bulk of the challenge. "It is far more difficult to penetrate a network, learn about it, reside on it forever and extract information from it without being detected than it is to go in and stomp around inside the network causing damage," said Michael V. Hayden, former director of both the NSA and the CIA who left office in 2009.[84]

The US-Israeli collaboration was intended to slow Iran's nuclear program, reduce the pressure for a conventional military attack, and extend the timetable for diplomacy and sanctions. Despite their collaboration on developing the malicious code, the United States and Israel have not always coordinated their attacks. Israel's April 2012 assaults on Iran's Ministry of Petroleum and oil-export facilities caused only minor disruptions. The episode led Iran to investigate and ultimately discover Flame. Some US intelligence officials were dismayed that Israel's unilateral incursion led to the discovery of the malware, prompting Iranian countermeasures. The disruptions led Iran to ask a Russian security company and a Hungarian cyber lab for help.[85] Researchers at the Kaspersky Lab, a Russian cybersecurity firm, reported their conclusion and named the malware "Flame."[86] Kaspersky Lab, which has clients around the world but is banned from US government facilities over concerns about its relationship to the Russian government, has identified Flame infections globally. While Flame was meant for Iran, the malware had spread to Israel and other Middle Eastern states, although not to Europe or North America. The infections have hit computers belonging to individuals, educational institutions, and state-related organizations.[87] The malware may have been in operation for as long as five to eight years before its discovery.

After it was eventually discovered, the creators of the malware sent a "suicide" command to remove it from most infected computers—Flame's creators did not have access to all the infected computers as security firms had won control of some of them. Symantec, for example, caught the command using booby-trapped computers set up to watch Flame's actions. Like many other security firms, Symantec has kept an eye on Flame using so-called "honeypot" computers that report what happens when they are infected with a malicious program. Symantec noticed that some Flame command-and-control computers sent an urgent command to the infected computers they were overseeing. The suicide command located every Flame file sitting on a computer, removed it, and then overwrote memory locations with gibberish to thwart forensic examination and to eliminate traces of the malware code. Analysis of the clean-up routine suggested it was written in early May 2012, said Symantec.[88]

Flame was used as a kick-starter to initiate the Stuxnet project.[89] "This is about preparing the battlefield for another type of covert action," said one former high-ranking US intelligence official, who added that Flame and Stuxnet were elements of a broader assault.[90] Findings reveal that the teams shared source code of at least one module between Stuxnet, Duqu, and Flame prior to 2010. The connection, Resource 207, was found in the 2009 version of Stuxnet, but was later removed from the 2010 version. Resource 207 has a lot in common with the code used in Flame, including the names of mutually exclusive objects, the algorithm used to decrypt strings, and similar approaches to file naming. Furthermore, the primary function of Resource 207 was to distribute the infection from one machine to another through removable USB drives.[91]

Other characteristics common to Flame, Stuxnet, and Duqu also suggest that the development teams on these operations were in contact with each other, including the ability of the malware to spread through computers that can share a printer on one network by exploiting a particular Windows vulnerability. Flame and Duqu are both cyberespionage malware, while Stuxnet was used for cyberattacks or physical destruction. However, analysts remain confident that the malware originated from completely different platforms used to develop multiple cyberweapons. Each of the specific malware codes has different architectures with different approaches that were used to infect systems and execute primary tasks, leading analysts to conclude that each piece of malware was separate and independent.[92]

Duqu

In addition to Flame and Stuxnet, there was another piece of malware discovered by researchers in the Laboratory of Cryptography and System Security at Budapest University of Technology and Economics. According to the laboratory that discovered and named it, "Duqu is not Stuxnet, but its structure and design philosophy are remarkably similar to those of Stuxnet. At this point in time, we do not know more about their relationship, but we believe that the creator of Duqu had access to the source code of Stuxnet."[93] In sum, the laboratory found the malware in the wild that had similarities to Stuxnet, including its modular structure, injection mechanisms, and a driver that had a fraudulent digital signature on it.[94] Open sources are not certain of the function of Duqu.

Gauss

Gauss has been linked to a suite of cyberweapons within Operation Olympic Games, including Flame, Stuxnet, and Duqu. Apparently, Gauss shares digital features that indicate they were made by the same developer. "After looking at Stuxnet, Duqu, and Flame, we can say with a high degree of certainty that Gauss comes from the same 'factory' or 'factories,'" states Kaspersky's analysis. "All these attack toolkits represent the high end of nation-state sponsored cyberespionage and cyberwar operations, pretty much defining the meaning of 'sophisticated malware.'"[95] Discovered in June 2012, Gauss has a main module that its creators named after the German mathematician Johann Carl Friedrich Gauss. Other components of the malware bear names of famous mathematicians, including Joseph-Louis Lagrange and Kurt Göde.

So far, Gauss is known to have infected between 2,500 and 10,000 computers—fewer than Stuxnet, but more than Flame and Duqu.[96]

Gauss appears to be designed to track the movement of funds between Iran and countries to which it may be clandestinely selling oil.[97] It does this by stealing detailed information from Lebanese bank computers (although it was also found on other machines) including browser history, cookies, passwords, and system configurations. Lebanese banks have served as clearinghouses for Iranian money. These stolen bank access credentials can be used to track the movement of other funds, too. For instance, Gauss' theft of credentials for various online banking systems and payment methods can also be used to detect if bank funds are moving from Iran and elsewhere to support the Syrian government.

In addition, Gauss targets users of Citibank and PayPal. While that might suggest that Gauss is crimeware, unlike most banking malware used by organized crime groups, "Gauss collects a lot of information about the host system, network information. It actually fingerprints the DNA of the computer it's on. . . . It's collecting reams of detailed information about the system that amounts to forensic proof for later legal prosecution or some other purpose. Criminal malware doesn't typically do this."[98] Importantly, embedded in Gauss is an encrypted payload reminiscent of Stuxnet, which is that it waits until it finds itself on precisely the correct system before it will activate.[99]

Cyber Disinformation Campaigns

Not all cyberespionage-cyberattacks require the use of malware. Shadow warfare has expanded the opportunities to magnify propaganda or, as it is more generally known, disinformation. While much cyber information is genuine information, there is also intentionally subversive information meant to cause harm to other states. It is difficult to have control of malicious information domestically, and even more difficult to determine if an international source is part of an intentional malicious disinformation campaign meant to confuse, retaliate, or cause social unrest. Certainly, the examples that most people are familiar with are propaganda campaigns that get into the mix of actual news stories and onto social media, whether intentionally or not. Cyber disinformation campaigns, like cyberespionage, are much easier to conduct in an interconnected world.

Cyber disinformation campaigns manipulate public opinion through false or misleading social media postings. They have become standard political practice across much of the world, from information ministries, specialized military units, and political operatives shaping the flow of information. These cyber disinformation campaigns exploit social media platforms like Facebook, Twitter, Instagram, and others. These efforts are often, though not always, clandestine, with the origin of the social media posts obscured. Social media platforms are infiltrated almost at inception by a range of international actors skilled at using information to advance political agendas, within their own countries and beyond.[100]

Cyber disinformation campaigns, to intensely amplify messages by automating the process of creating and delivering posts, rely on human users and computerized bots. A bot is a piece of software that carries out an automated internet task, often performing simple repetitive tasks such as browsing the internet for information similar to search engines or, maliciously, launches denial of service attacks, harvests email addresses, or scrapes content and manipulates comments or votes on sites. Bots interact with human users and also with other bots. They often play key roles by automatically creating social media posts, responding to other users, and echoing select themes in a way that are difficult to distinguish from ordinary human users. Bots can post far more often than human users, in some cases more than a thousand times a day. Human users, dubbed "cyborgs," rely on similar automation technology to bolster the power of their accounts. Cyber disinformation campaigns through human users and bots are issuing false news reports, attacking journalists, or supporting a government position or political view.[101]

One of the most notorious disinformation campaigns includes the Russian support for Brexit in the UK. There have been a series of UK inquiries into Russian interference, most notably the House of Commons' *Disinformation and 'fake news': Final Report Eighth Report of Session 2017–19 Report*, stating

> We repeat our call to the Government to make a statement about how many investigations are currently being carried out into Russian interference in UK politics. We further recommend that the Government launches an independent investigation into past elections—including the UK election of 2017, the UK Referendum of 2016, and the Scottish Referendum of 2014—to explore what actually happened with regard to foreign influence, disinformation, funding, voter manipulation, and the sharing of data, so that appropriate changes to the law can be made and lessons can be learnt for future elections and referenda.[102]

Various investigations,[103] including the above UK government committee report, question the founder of Leave.eu, British businessman and political donor Aaron Banks' alleged connections to both Russian money and Cambridge Analytica (and its parent company Aggregated IQ), the shadowy political consulting firm that became infamous during the 2016 US presidential election, in part because it harvested Facebook user content without consent.

It is hard to know how these disinformation campaigns affected election outcomes. However, the University of Edinburgh research indicates that more than 400 Russian-run Twitter accounts active in the 2016 US election were also actively posting about Brexit. In addition, University of California at Berkeley and Swansea University researchers identified 150,000 Twitter accounts with Russian ties that disseminated messages about Brexit.[104] What can be said, in general, about the links between foreign money, disinformation, and hidden agendas is that they can cause social unrest and magnify social divisions in liberal democracies.

POLICY PATTERNS

What all these cyberattacks, cyberweapons, and cyber campaigns reveal is that not only is shadow warfare real, but that it has been developing in sophistication and frequency over the past few decades. As shadow warfare develops into continuous warfare and the dissolution of the sharp distinctions between allies and adversaries, fierce debates over how to regulate or develop national and international policies to cope with this new warfare have erupted. These elements of shadow warfare are illustrated by the major powers struggles in the following chapters.

Not surprisingly, major cyber powers—especially the United States, Russia, and China—have developed nuanced policy decision-making institutions to wage this type of warfare. Most of these developed out of older institutions. Military institutions, intelligence agencies, and departments that used to look at cryptology and weapons development were tasked with the original shadow war functions. Over the past three decades, these institutions have morphed into more task-appropriate agencies, although they are still mostly led or housed within the military and intelligence agencies of the great powers.

More surprisingly, these policies tend to be controlled by the very top of the great-power governments. In the United States, specific cyberattack targets can be chosen by the president. In Russia, the president directs plans for specific targets and campaigns. In China, the president has reorganized institutions in order to have more direct control over cyberattacks and cyberwar planning. At some stage, privateers were recruited in the United States, Russia, and China to participate in public campaigns of DDoS or hacktivism. But those days are gone. In some ways, there are tinges of the thirteenth- to eighteenth-century wars as games by princes, of which people were rarely told, and even more rarely consulted. Even within their own countries, shadow warfare is fought outside of much public awareness.

New cyber strategies have arisen out of these newly developed institutions and leadership. The nature of these strategies varies significantly between the great powers. Where the United States uses complicated and stealth cyberweapons that are designed for specific purposes—and now mostly self-destruct when the objective is obtained—in order to force specific actions or inactions on the world stage, Russia relies on a more hit-and-miss strategy that uses massive disinformation campaigns. As opposed to the specific targeting by the United States, Russia looks to broadly undermine the capabilities of liberal democracies and to widen existing social rifts in order to weaken the overall society and hence the target country's power and prestige through disinformation. China's strategy is primarily based on building its economy in order to increase its overall power projection. Thus, the strategies that China uses are based on gathering economic, technological, and military developments through extensive cyberespionage. All three great cyber powers have strategies that are primarily military in concept, focusing on taking down adversaries' infrastructure and incapacitating major urban centers. All three powers use cyberwarfare as a permanent form of warfare.

Cyberattacks are one measure of a country's cyber strategy and a test of a country's cyberwarfare institutions. The US cyberattacks are far-reaching. Not only has the United States used cyberattacks against Russia for decades, it is now also using intense cyberattacks and cyberespionage against China. As the United States maintains its role as global policeman, other countries, too, become targets of cyberattacks, like the United States trying to delay North Korea's missile testing or disable the centrifuges used in Iran's nuclear power program. And as the revelations made by former NSA contractor Edward Snowden and others have shown over the past years, the United States uses a globally intensive program of cyberespionage that includes allies and adversaries alike.

Russian cyberattacks are focused on its near neighbors, especially those that were once part of the Soviet Union but have become outspoken critics, like Estonia, Georgia, and Ukraine. These Russian attacks often combine disinformation campaigns with cyberattacks on parts of a country's infrastructure like Estonia's banking system or Ukraine's electric grid. Russian attacks on Europe and the United States are often launched in order to specifically influence political elections and to cause underlying damage to domestic opinion. China, too, uses cyberattacks to address specific regional issues, such as promoting its claims—as well as undermining the claims made by other nations—to the South China Sea. It also engages in extensive cyberattacks on neighboring powers like Australia and India, and persistently conducts attacks on Taiwan and the island's political process.

As the analysis of US, Russian, and Chinese policies indicates, all three powers conduct themselves as if cyberspace can be controlled. The United

States has a multi-decade program that maps the internet, follows occurring cyberattacks, and traces the origins of specific cyberweapons and malware back to their owners in an attempt to impose a deep understanding of cyberspace. With much, although clearly not all, of cyberspace originating in the United States, home to organizations like the Internet Corporation for Assigned Names and Numbers (ICANN) that provide a modicum of regulation on the internet, the United States continues to act as if it owns and, therefore, controls cyberspace. The Russian and Chinese stances are different from the US position. China, from the inception of the internet in its country, has attempted to control cyberspace by creating its own national intranet and by creating walls and barriers to keep ideas, cyberweapons, and cyberespionage out. China has used the uniqueness of the Chinese language to help control what happens domestically on its intranet. While this has not been a complete success, it has allowed Beijing to gain a strong sense of control over cyberspace. Russia has had a different experience. While the Soviet Union was a world leader in mathematics, natural science, and computing, Russia was not able to continue those feats after the collapse of the USSR. The current projection of control over cyberspace is based on a belief that it can regain that earlier footing and a belief that it can institute a cyberspace nationalism similar to China's intranet, albeit from a vastly different starting point.

The projection of shadow warfare is that it is silent and undetectable. Scholarship insists that it be analyzed and critiqued. The following chapters offer some insight into the cyber policies of the United States, Russia, and China. This war is knowable. It should be open to discussion in order that larger swaths of societies help direct this new warfare that is shaping many aspects of the world.

Chapter 2

Cyber United States

As shadow warfare becomes increasingly relevant in calculations of power, power projection, security, and strategic planning, it is not surprising that the three big players in shadow warfare are the three big players in many aspects of security: the United States, Russia, and China. This chapter looks at the national policies and the underlying doctrines of the United States in the era of shadow warfare as well as examples of cyberattacks, such as who is being attacked, how, and why. The following chapters address the issues of shadow warfare for Russia and China.

The major players all view cyber operations as clandestine and are reluctant to discuss their shadow warfare strategies and tactics in other than general terms, so investigative journalism, strong inferences, and broad government documents must be the guide.[1] Fortunately, cyber signatures can be forensically tracked with a high degree of confidence, with the media increasingly reporting on more and more incidents uncovered by journalists. There are some commonalities among the three major players. First, all participate in cyberespionage. Second, all believe they must have strong, albeit different, cryptanalytic programs.[2] Third, all believe that cyberspace is controllable. Fourth, all have conducted cyberattacks. Fifth, both the United States and Russia use cyberattacks as part of a hybrid war effort, and China has incorporated hybrid war into its planning. Sixth, although the United States and China once made significant use of privateers as part of their cyber strategy, Russia still does. Finally, and most importantly, all three major players accept a state of perpetual war through shadow warfare as a natural condition.[3]

For at least two decades, since the 1990s, the United States was at the forefront of new types and techniques of shadow warfare, especially cyberwarfare. In some respects, it continues to lead. In other areas, it has peers with different approaches. Shadow warfare has become a significant factor in US strategy in addition to, and in combination with, traditional and nuclear strategies. The US position is succinctly wrapped up in the following quote from its National Security Strategy: "America's response to the challenges and opportunities of the cyber era will determine our future prosperity and security."[4]

The 2017 National Security Strategy addresses the importance of cyberwarfare to US strategy in both defensive and offensive terms. Defensive strategy is discussed first and extensively by stating:

> For most of our history, the United States has been able to protect the homeland by controlling its land, air, space, and maritime domains. Today, cyberspace offers state and non-state actors the ability to wage campaigns against American political, economic, and security interests without ever physically crossing our borders. Cyberattacks offer adversaries low cost and deniable opportunities to seriously damage or disrupt critical infrastructure, cripple American businesses, weaken our federal networks, and attack the tools and devices that Americans use every day to communicate and conduct business. Critical infrastructure keeps our food fresh, our houses warm, our trade flowing, and our citizens productive and safe. The vulnerability of U.S. critical infrastructure to cyber, physical, and electromagnetic attacks means that adversaries could disrupt military command and control, banking and financial operations, the electrical grid, and means of communication.[5]

Offensive strategy is mentioned as well but less so, and in vaguer terms, saying that "cyber operations against adversaries can be conducted as required."[6]

US shadow deterrence strategy, also outlined in the 2017 National Security Strategy, focuses on the need to make the cost of cyber operations against the United States too costly for attackers to undertake:

> Cyberattacks have become a key feature of modern conflict. The United States will deter, defend, and when necessary defeat malicious actors who use cyberspace capabilities against the United States. When faced with the opportunity to act against malicious actors in cyberspace, the United States will be risk informed, but not risk averse, in considering our options. We will invest in capabilities to support and improve our ability to attribute cyberattacks, to allow for rapid response. We will improve our cyber tools across the spectrum of conflict to protect U.S. Government assets and U.S. critical infrastructure, and to protect the integrity of data and information.[7]

The US Department of Homeland Security updated its Cyber Security Strategy for civilian government purposes in May 2018, citing evolving threats from cyberspace that the state is facing. The document lists the strategy's five pillars, focusing on identifying risks, reducing vulnerability, reducing threats, mitigating consequences, and enabling cybersecurity outcomes.

INSTITUTIONS AND INDIVIDUALS

Several US institutions bear responsibility for cyber activities. Responsibilities are divided between the Department of Homeland Security (DHS), the Federal Bureau of Investigation (FBI) at the Department of Justice, the Central Intelligence Agency (CIA), the National Security Agency (NSA), and the Department of Defense (DoD), including the Strategic Command's Cyber Command.[8] DHS has the primary defensive role for the US government, coordinating domestic defense. Offensive operations are most likely assigned to Cyber Command and to elements of the CIA. Coordinating between and among federal agencies in response to a significant cyber incident is the Cyber Threat Intelligence Integration Center (CTIIC). It is a small, multiagency center within the Office of National Intelligence that works to increase the speed at which the US government recognizes that cyber activity is threatened or occurring, and whose mission is to provide integrated all-source analysis of intelligence related to foreign cyber threats or incidents affecting US national interests. Presidential Policy Directive 41 on Cyber Incident Coordination names CTIIC as one of the three federal lead agencies (with the DHS and the FBI) to coordinate the response to a significant cyber incident.[9]

Tasked with primary responsibility for domestic defense, the DHS's National Cyber Security Division works "collaboratively with public, private, and international entities to secure cyberspace and America's cyber interest."[10] The division also has a number of programs to protect cyber infrastructure from attack.[11] Operating under the National Cyber Security Division, the National Cyber Response Coordination Group is comprised of thirteen federal agencies and is responsible for coordinating the federal response in the event of a nationally significant cyber incident.[12] The Cyberspace Review Policy, completed in 2011, outlines the roles of federal agencies to secure cyber infrastructure.[13]

DoD's Cyber Command, one of ten unified commands in the military, is responsible for dealing with threats to the military cyber infrastructure. The Pentagon raised the US Cyber Command to the status of a unified combatant command on May 4, 2018, after eight years of it working as a sub-unified organizational unit under the US Strategic Command. Now, as a separate unified body Cyber Command can smoothly leverage the technical expertise of the NSA and the Defense Advanced Research Project Agency (DARPA). DARPA is a Pentagon division that focuses on experimental efforts. Cyber Command has three missions: day-to-day protecting of all defense networks, establishing a single chain of command running up to the president, and working with various partners to share threat information and help coordinate responses.

Cyber Command has extensive offensive capabilities for breaking into and destroying foreign communications and computer networks. Cyber Command's service elements include Army Forces Cyber Command, the Twenty-Fourth Air Force, Fleet Cyber Command, and Marine Forces Cyber Command.[14] US Cyber Command's 133 teams are at full operational capability, meeting a rigorous set of criteria, including an approved concept of operation and trained personnel; the focus will shift toward readiness to perform the mission and deliver optimized mission outcomes, continuously.[15] Military cyber operations, however, are constrained by governing legal authorities. Military cyber operations that result in the disruption, destruction, or manipulation must be approved by the US president. Indicating that the rules for responding in an escalated manner in cyberspace, or with conventional retaliation, would require decisions by the civilian leadership, General Martin Dempsey, who served as the eighteenth chairman of the Joint Chiefs of Staff, said that "if it became something more widespread and we needed to do something beyond that, it would require interagency consultation and authorities at a higher level in order to do it."[16] These constraints largely do not apply to the NSA or the CIA.

The NSA and the CIA also have the capability to offensively break into foreign computer networks, as well as gather information. It was the NSA's programs that were leaked in 2013 by Edward Snowden, an NSA contractor and former CIA employee. More recently, in the summer of 2016, a set of stolen NSA cyber tools from its arsenal for penetrating foreign computer networks were auctioned on the web by a group calling itself the Shadow Brokers.[17]

The NSA, along with its electronic eavesdropping and code-breaking capabilities, develops cyberattacks aimed at US adversaries—and sometimes allies. The NSA's Office of Tailored Access Operations (TAO) has almost 1,000 operators and support staff working around the clock on rotating shifts. TAO's operations include stealing passwords, data, and text messages and analyzing foreign communication infrastructure for weaknesses that could be exploited by cyberweapons.[18] The CIA[19] may not have the NSA's sophistication in building malware, but it is deeply involved in cyber operations.[20] Within the CIA, there has been a major expansion of the Information Operations Center (IOC). The IOC is one of the CIA's largest divisions, employing hundreds of people at facilities in northern Virginia. Its primary focus—once counterterrorism—is now cybersecurity. The IOC undertakes offensive operations as well as the recruitment of new intelligence sources.[21] The IOC's annual cyberwar exercise, Silent Horizon, has been taking place since 2007.[22] Both the NSA and the CIA analyze the intelligence obtained and continue to develop new weapons even as recent attacks, such as those on North Korea and Iran, have been exposed.[23]

Questions arise in regard to cyberweapons that are captured "in the wild," purchased from disgruntled employees, or traded by weapons brokers. Gil Baram outlines these concerns in his article for the *Council on Foreign Relations*, using North Korea's WannaCry and Russia's NotPetya as examples of where US-developed tools were reused on the cheap.[24] As Baram discusses, when a major cyber actor like the United States invests in developing offensive cyberweapons that are then stolen and reused, it raises critical questions of urgent policy relevance. Unlike most weapons, cyberweapons are not necessarily destroyed during use and can be captured and applied by the intended targets at a lower cost than developing their own. Additionally, the theft and reuse of cyberweapons have changed the way the United States handles vulnerabilities that leak into the open, including developing information-sharing mechanisms to address it through a policy titled Vulnerabilities Equities Policy and Process.[25] The policy states that "the primary focus of this policy is to prioritize the public's interest in cybersecurity and to protect core Internet infrastructure, information systems, critical infrastructure systems, and the U.S. economy through the disclosure of vulnerabilities discovered by the USG, absent a demonstrable, overriding interest in the use of the vulnerability for lawful intelligence, law enforcement, or national security purposes."[26]

The protection of the US government and domestic structures is the responsibility of the DHS, while corporate infrastructure is the responsibility of private companies.[27] One problem for highly developed countries like the United States is that much of the critical infrastructure is private. General Keith B. Alexander, who served as director of the NSA and as the first head of Cyber Command, noted that Cyber Command is trying to determine whether such activities as commercial espionage or theft of intellectual property are criminal activities or "breaches of national security,"[28] as opposed to straightforward governmental cyberespionage, which is clearly a national security issue. The DHS's protection of the digital infrastructure of nonmilitary government sectors comes under the National Cybersecurity and Communications Integration Center's CERT. CERT defends against cyberattacks within the "dot gov" domain and is responsible for security collaborations with both the government and private industry. The DHS has identified seventeen sectors of US critical infrastructure that must be protected, including the defense industrial base, financial systems, transportation networks, and water works. The DHS and the DoD signed a cybersecurity pact in September 2010 formalizing their cooperation, allowing the colocation of personnel, joint operational planning, and allowing the DHS to use the NSA's advanced technical expertise.[29]

The Comprehensive National Cybersecurity Initiative (CNCI), launched by US president George W. Bush in January 2008, was expanded by President Barack Obama to become key in a broader, updated national US cybersecurity strategy. These CNCI initiatives played a key role in supporting the US Cyberspace Policy Review, including creating or enhancing shared situational awareness of network vulnerabilities, threats, and events within the US government, enhancing US counterintelligence capabilities, and expanding cyber education, as well as working to develop strategies to deter hostile or malicious activity in cyberspace. Additionally, the CNCI includes funding within the federal law enforcement, intelligence, and defense communities to enhance criminal investigation, intelligence collection, processing, analysis, and information assurance critical to enabling national cybersecurity efforts.[30]

CYBER STRATEGY

The US cyber strategy incorporates the usual rationales of protecting US national interests and global standing. The overarching strategy rests on an understanding that, more than many other states, the United States is militarily, economically, and socially dependent on cyber institutions, ranging from global positioning systems to electronic banking to social networks. Moreover, the US cyber strategy is designed for limiting US casualties, maintaining military dominance, supporting international trade, and promoting American values.

The US cyber strategy is stated in the 2018 US Cyber Command's Command Vision, entitled "Achieve and Maintain Cyberspace Superiority."[31] This document recognizes that "adversary behavior [is] intentionally set below the threshold of armed aggression [for] strategic effect."[32] This strategy focuses on adversarial cyber operations as well-thought-out campaigns seeking to avoid significant American reaction while degrading US power and advancing the attacker's own relative capacities. Thus, the US cyber strategy recognizes that cyber operations are a new arm in the distribution of power and can impact relative power without traditional armed aggression.[33]

A second element in the 2018 Command Vision is the recognition that the United States "faces peer competitors in the cyberspace domain."[34] This is a relatively new recognition by the United States, which was fortunate to have initial superiority in the cyber realm. The low cost and high impact of cyberweapons and cyberespionage on the global stage have seen the United States lose some of its early advantages, most notably to the other two great powers—Russia and China.

Finally, the 2018 Command Vision acknowledges that the norms of cyberspace are somewhat chaotic, negative, and operate without real constraint below the threshold of war. A new approach that the US Cyber Command offers is to take more offensive actions as a defensive strategy—sometimes known as active defense—increasingly militarizing cyberspace. The United States explicitly blames the militarization of cyberspace on the actions of adversaries, although the United States itself was an early participant in this militarization. Similarly, the United States emphasizes that the Command Vision is not an offensive doctrine but a seamless operational approach integrating resilience and defense against adversarial activity. Nonetheless, cyberweapons are continually being developed within the United States for cyber operations.[35]

The current US cyber strategy rests upon decades of earlier US cyber policy formulations. The United States was fundamental in the development of the internet and took an early role protecting itself from cyberattacks and using cyber tools to attack others. Initially designed by DARPA with funding from the National Science Foundation (NSF), the internet was originally used for academic, scientific, and economic purposes. In the final years of the twentieth century, President Bill Clinton's 1995 National Security Strategy mentioned a "threat of intrusion to our military and commercial information systems."[36] In May 1998, the Clinton administration issued a presidential policy directive to warn of the dangers of potential cyberattacks on the county's vital infrastructure and called for a national cyberspace protection plan by 2000.[37]

Cyberspace was first officially suggested for military operationalization during George W. Bush's time in office, in the 2004 National Military Strategy and then in the March 2005 National Defense Strategy, which identified cyberspace as a new theater of operations and assessed cyberspace operations as a potentially disruptive challenge.[38] This was further developed during the Obama administration under the stated goal that

> the United States will work internationally to promote an open, interoperable, secure, and reliable information and communications infrastructure that supports international trade and commerce, strengthens international security, and fosters free expression and innovation. To achieve that goal, we will build and sustain an environment in which norms of responsible behavior guide states' actions, sustain partnerships, and support the rule of law in cyberspace.[39]

The Trump administration's first serious foray[40] into US cyber policy was the 2018 Command Vision. This vision sees the United States as the dominant power in cyberspace and includes the use of offensive weapons as a clear aspect of policy.

In developing US strategy for cyberspace, the DoD focuses on a number of central aspects of the cyber threat, both offensive and defensive—external threat actors, insider threats, supply chain vulnerabilities, and threats to the DoD's operational ability—first in the 2009 Cyberspace Review Policy, the 2010 National Security Strategy,[41] and then the 2011 Strategy for Operating in Cyberspace.[42] The May 2011 International Strategy for Cyberspace states that the United States "reserves the right to use all necessary means" to defend itself and its allies and partners, but that it will "exhaust all options before [the use of] military force."[43] The 2015 DoD Cyber Strategy expanded on its strategic goals to include international partnerships: build and maintain ready forces and capabilities to conduct cyberspace operations; defend the DoD information network, secure DoD data, and mitigate risks to DoD missions; be prepared to defend the US homeland and US vital interests from disruptive or destructive cyberattacks of significant consequence; build and maintain viable cyber options and plan to use those options to control conflict escalation and to shape the conflict environment at all stages; and build and maintain robust international alliances and partnerships to deter shared threats and increase international security and stability.[44] This has now shifted to a more offense-based policy in the Trump administration, which has not allowed members of Congress to read a classified directive President Trump issued in 2018 outlining new rules for the military's use of cyberweapons to increasingly deploy offensive cyber operations against adversaries, including against Iran in June 2019.[45]

There are several components in the US cyber strategy. One is for gathering intelligence. The United States uses cyber means and other technological methods to obtain intelligence both inside and outside the state. Although the main force of intelligence gathering is now done through cyber operations, traditional methods of intelligence gathering are still used. The United States asserts that it does gather intelligence for governmental—but not for commercial or financial—purposes.[46] Nonetheless, although it is impossible to verify through open sources, it seems clear that the United States does large-scale intelligence gathering through cyberspace.[47]

A second component within the US cyber strategy is for defensive purposes. The DoD Defense Science Board Task Force on Cyber Deterrence states that "the United States gains tremendous economic, social, and military advantages from cyberspace. However, our pursuit of these advantages has created extensive dependencies on highly vulnerable information technologies and industrial control systems."[48] Many US command and control, military weapons, and communications systems rely heavily on cyber connectivity. With a presumption of a possible breach of defensive barriers, the DoD continues to develop resilient networks and systems as well as remain operationally effective by isolating and neutralizing the impact, using redundant capacity, or shifting its operations from one system to another. Moreover, the DoD is identifying options for shifting its operations to secure networks.[49] The Defense Science Board Task Force on Cyber Deterrence report identifies Russia and China, as well as a few other states, as major threats to US cyber defense.[50]

A third component within the US cyber strategy is for offensive purposes. The United States, like most cyber powers, was initially reluctant to acknowledge its use of offensive cyberweapons. Although military officials insisted that their cyber strategy remained defensive for a decade to preserve the international norm against acts of aggression,[51] the move to offensive operations was revealed in 2012. In August that year, the US Air Force signaled readiness to go on the cyber offensive, announcing that it was looking for ideas on how "to destroy, deny, degrade, disrupt, deceive, corrupt, or usurp the adversaries [*sic*] ability to use the cyberspace domain for his advantage."[52] The emphasis on offensive operations was confirmed by a program launched in October 2012 by DARPA,[53] the Pentagon's emerging technologies research agency. DARPA's new program, dubbed Plan X, was "to create revolutionary technologies for understanding, planning, and managing cyberwarfare."[54] The acknowledged expansion into offensive operations represented an evolution in cyber strategy, partly due to US economic and military power being heavily dependent on technology[55] and partly due to a technological leap in offensive cyberweapons themselves.

This third, offensive component has two major parts. First is the use of cyber offensives to replace kinetic or traditional warfare. One classic example of stand-alone cyberattacks occurred in June 2012, when it was revealed that the United States and Israel were behind the Stuxnet attack on Iran. Second is the use of cyber offensives in conjunction with traditional warfare, referred to as hybrid warfare. A hybrid war strategy includes a multilayered effort—with both kinetic (conventional physical military) elements and cyber components—which is designed to negatively impact military defenses, like command and control, as well as to socially destabilize and polarize a state by influencing policymakers and the population.[56]

In purely practical terms, in hybrid war, operational commanders attack an opponent's computer and information systems while protecting their own information and communication networks. For instance, electronic penetrations have preceded conventional military attacks, such as disabling Iraq's military computers before the US invasion in 2003.[57] In another instance, a former ground commander in Afghanistan, US Marine lieutenant general Richard Mills, acknowledged using cyberattacks while directing international forces in southwest Afghanistan in 2010, stating, "I was able to use my cyber operations against my adversary with great impact."[58]

In American hybrid warfare, socially influencing populations and policymakers through information operations or disinformation campaigns have long been referred to as psychological operations (PSYOPs), or occasionally as military information support operations[59] or political warfare. While newspaper articles and pamphlets were traditionally used—and are still used—in non-online communities, the cyber domain is actively exploited by the United States to conduct influence operations via cell phones, emails, text messages, and blogs in both peacetime and combat environments.[60] Political warfare was defined in 1948 by US diplomat George Kennan: "Political warfare is the logical application of Clausewitz's doctrine in time of peace. In broadest definition, political warfare is the employment of all the means at a nation's command, short of war, . . . [ranging] from . . . 'white' propaganda to such covert operations as clandestine support of 'friendly' foreign elements, 'black' psychological warfare and even encouragement of underground resistance in hostile states."[61] Examples of information operations include text messages delivered to cell phones or even to specific cell phone towers to enable regular news updates to a target audience. Disinformation operations include social networking sites containing propaganda videos, doctored photos or slanted news stories, and memes.[62] Much of the work is carried out by military information support teams that operate multilingual news sites tailored to specific regions, like the *Southeast European Times* for the Balkans and the *Magharebia* for North and West Africa.[63]

The other two major US efforts in cyberspace other than gathering intelligence include defending computer networks and carrying out offensive attacks. However, the line between defense and offense can be blurry. For instance, in the case of a foreign cyberattack on US infrastructure, *The New York Times* quotes from a 2013 speech by General Martin Dempsey that "our first instinct will be to pull up the drawbridge and prevent the attack, that is to say, block or defend."[64] If the cyberattack could not be repulsed, the next response is "active defense," which General Dempsey defined as a "proportional" effort "to go out and disable the particular botnet that was attacking us."[65] The comments signal that the United States is redefining defense as requiring an active defense capacity to reach forward over computer networks and take preemptive action, blurring the line between offense and defense if the United States detects or suspects a threat. Similarly, offensive measures could be used in a punishing response for a first strike cyberattack on a US target.[66]

A fourth component within the US cyber strategy is for deterrence purposes. Even more complicated than cyber defense or offense, cyber deterrence is discouraging an act by instilling doubt or fear of the consequences. Deterrence too has two parts: deterrence by denial and deterrence by cost imposition. Deterrence by cost imposition has a different balance depending on the perpetrator and the severity of the attack to be deterred. There are doubts concerning the effectiveness of cyber deterrence—it is difficult to measure because it is the art of what did not happen. The then-director of

National Intelligence James Clapper argued in his testimony to the Senate Armed Services Committee on January 5, 2017: "We currently cannot put a lot of stock, at least in my mind, in cyber deterrence. Unlike nuclear weapons, cyber capabilities are difficult to see and evaluate and are ephemeral. It is accordingly very hard to create the substance and psychology of deterrence in my view."[67] Cyber deterrence can occur by the elimination of detected malware. Cyber retaliation—as well as non-cyber retaliation for cyberattacks—as part of cyber deterrence certainly seems to have a role in the US cyber strategy. For instance, the US government approved a covert cyber deterrence measure in retaliation for Russian interference in the 2016 US presidential election by implanting computer code, which Russia was bound to find, in sensitive computer systems in order to remind Moscow of America's cyber reach.[68] Given its objections to additional punitive measures by the United States, Russia seemed to have located the codes.

The issues with cyber deterrence have led the DoD Task Force on Cyber Deterrence to argue that it is important to have credible non-cyber as well as cyber responses to cyberattacks.[69] Moreover, the DoD Task Force on Cyber Deterrence argues the United States should clarify that it will respond to all cyberattacks and other costly cyber intrusions by well-outlined norms and impose costs exceeding any possible benefit for potential attackers.[70]

CYBERESPIONAGE

Cyberespionage is increasingly merging with cyberattacks and cyberwarfare in the United States as well as in some other states. Cyberespionage has dramatically increased the role of espionage in the era of shadow warfare. This has raised the issue of espionage—as well as the use of cyberespionage as the first level of attack before cyberweapons are employed—to new levels. Cyberespionage in all three of the great powers exists both domestically and internationally.

Domestic cyberespionage is allegedly part of the shadow strategy of monitoring terrorist activity inside national bonders, by both citizens and noncitizens. Domestic cyberespionage must meet—sometimes succeeding or failing—a balance between public privacy and public safety in a liberal democracy. The US battle between intelligence and public privacy is illustrated by the early example of the 1961 Bay of Pigs fiasco, a classic tale of how CIA overconfidence, combined with presidential inexperience, led to a wildly flawed policy in the goal of safety. The intent of the Bay of Pigs operation was to covertly overthrow the newly installed government of Fidel Castro in Cuba, but the ramifications were the overt collapse of the intelligence effort itself, a strengthening of the Cuban position, and a decrease in US public safety as a result of the Cuban Missile Crisis that followed in 1962.[71]

Another program—the NSA's Project SHAMROCK, an early foray into domestic cyberespionage—erred strongly on the side of security over privacy shortly after World War II, persuading three major American telegraph companies to hand over most of their traffic. By the time the program was shut down in 1975, the NSA had collected information on some 75,000 American citizens, especially those active in the antiwar movement,[72] sharing its information with the CIA. The CIA was running its own illegal domestic intelligence program called Operation CHAOS. In 1978, Congress, in an attempt to restore the balance between privacy and safety, created the 1978 Foreign Intelligence Surveillance Act, or FISA, which forbade the intelligence agencies to spy on anyone in the United States without probable cause to believe that the person was an agent of a foreign power. In 1999, during the Clinton administration, intelligence agencies were desperate to discover links between al-Qaeda operatives and potential terrorists in the United States. The NSA collected a trove of telephone metadata, but the Justice Department advised the NSA that the plan was tantamount to illegal electronic surveillance.[73]

A long-term shift in the privacy-safety balance in US domestic cyberespionage occurred after the September 11, 2001, attacks, and the resulting Patriot Act. The George W. Bush administration created Stellar Wind, four phone and internet-surveillance programs, including two programs that collected the content of emails and phone calls, and two metadata programs. This domestic cyberespionage continued under the Obama administration from 2008 onward in order to find the domestic contacts of potential terrorists. The legal case for phone and internet content collection was harder to make than the arguments concerning metadata since the Supreme Court had ruled in 1979 that metadata was not covered by the Fourth Amendment to the US Constitution, but the content of phone calls and emails was. By 2011, the NSA's email and phone content programs were collecting domestic communications of tens of thousands of Americans. Yet three of the four original Stellar Wind programs—the phone-metadata program and the content-collection programs—have expanded and are still running,[74] now under the FISA court supervision following a dramatic hospital bed confrontation over the legal issues between the Department of Justice and the White House.[75] In March 2004, after a section of the Justice Department concluded the email program was not legal, then-acting attorney general James Comey refused to reauthorize it. As *The Washington Post* reported at the time, "That refusal resulted in a dramatic showdown that month between Attorney General John Ashcroft, who was in the hospital with a severe pancreatic ailment, and White House counsel Alberto Gonzales, who had rushed to Ashcroft's hospital bedside in a futile attempt to persuade him to reauthorize the e-mail program."[76] Nonetheless, while US legal rationales have shifted over time, some surveillance programs have become even broader and more intrusive than the original Stellar Wind.[77]

In January 2018, the Trump administration signed into law a bill extending Section 702 of FISA that renewed the NSA's warrantless internet-surveillance program. Under the law, the NSA eavesdrops on vast amounts of digital communications from foreigners living outside the United States via companies like Facebook, Verizon, and Google. The program also intercepts US citizens' communications, including when they communicate with a foreign target living overseas, and can search those messages without a warrant.[78]

While domestic cyberespionage continues, the main force of the US espionage effort is foreign cyberespionage, which it views as a legitimate activity, albeit with consequences if caught.[79] While government-built malware for gathering intelligence is routinely used around the world, the United States distinguishes between intelligence gathering for government purposes and intelligence gathering for commercial practices and financial gain. According to an e-mailed statement from a DoD NSA spokesperson, "The Department of Defense does engage" in computer network exploitation, clarifying that "The department does ***not*** engage in economic espionage in any domain, including cyber."[80]

Not all major powers agree with the US distinction between government and economic espionage. Most notably, Chinese officials challenge the US position. The United States does conduct a vibrant cyberespionage campaign against China, which ranks high on Washington's priority list. The NSA used cyberespionage to enter computers belonging to Huawei and China Telecom, which became public when Edward Snowden leaked documents evidencing US cyberespionage.[81] The United States justifies economic espionage. Jack Goldsmith, former assistant attorney general for the DOJ's Office of Legal Counsel from 2003 to 2004 and DoD special counsel, quotes former CIA director Stansfield Turner saying in 1991 that "as we increase emphasis on securing economic intelligence, . . . we will have to spy on the more developed countries—our allies and friends with whom we compete economically."[82] He also quotes former CIA director James Woolsey confirming in 2000 that the United States steals economic secrets from foreign firms and their governments "with espionage, with communications, with reconnaissance satellites."[83] And, finally, Mr. Goldsmith also quotes former director of National Intelligence James Clapper in 2013 stating, "What we do not do . . . is use our foreign intelligence capabilities to steal the trade secrets of foreign companies on behalf of—or give intelligence we collect to—US companies to enhance their international competitiveness or increase their bottom line."[84] Nonetheless, this permits economic espionage of foreign governments and institutions, and the theft of trade secrets from foreign firms—just not "on behalf of" US firms.[85]

However, the United States asserts that this cyber intelligence gathering is a standard activity in the game of international espionage. While the US government does not like China using cyberespionage for normal intelligence gathering, US officials reserve their ire for state-sponsored intellectual-property theft, which will be discussed in more detail in the next chapter.[86] China seems to be at least somewhat aware of the distinction the United States is making. Neither state wants to discuss military espionage.[87]

CYBERATTACKS

The United States has been a primary purveyor of cyberattacks in ongoing continuous shadow warfare. After US intelligence services carried out 231 offensive cyber operations in 2011,[88] President Barack Obama ordered a list of potential overseas targets for US cyberattacks. An eighteen-page Presidential Policy Directive 20, issued in October 2012, stated that Offensive Cyber Effects Operations (OCEO) "can offer unique and unconventional capabilities to advance US national objectives around the world with little or no warning to the adversary or target and with potential effects ranging from subtle to severely damaging." Presidential Policy Directive 20 defines OCEO as "operations and related programs or activities . . . conducted by or on behalf of the United States Government . . . to enable or produce cyber effects outside United States government networks."[89] Of the 231 offensive operations conducted in 2011, almost three-quarters were against top-priority targets, including adversaries such as Iran, Russia, China, and North Korea.[90] It is reasonable to project that the quantity and quality of cyberattacks emanating from the United States have significantly increased since 2011.

In 2017, the DoD Defense Science Board Task Force on Cyber Deterrence defined a cyberattack in US parlance as "any deliberate action that affects the desired availability and/or integrity of data or information systems integral to operational outcomes of a given organization."[91] These cyberattacks attempt to be covert. For instance, Umbrage is a voluminous library of cyberattack techniques that the CIA collected from malware produced by other countries, including Russia, which allows the CIA to mask the origin of some of its cyberattacks in an attempt to confuse forensic investigators.[92] In another instance, under an extensive effort, code-named Genie, US cyberattackers enter foreign networks in order to put them under surreptitious US control. The $652 million project has placed "covert implants"—remote sophisticated malware—in computers, routers, and firewalls on tens of thousands of machines every year, with plans to expand into the millions.[93] Keeping cyberattacks covert is becoming increasingly difficult as top cybersecurity firms join cyber operatives in the major states to decrypt and identify originators.

The NSA knows its decryption abilities are important, saying that "it is the price of admission for the US to maintain unrestricted access to and use of cyberspace."[94] The full extent of the NSA's decoding capabilities is known only to a limited group of top analysts from the so-called Five Eyes: the United States, the United Kingdom, Canada, Australia, and New Zealand, with a parallel British Government Communications Headquarters (GCHQ) counter-encryption program. Unlike some classified information that can be parceled out on a strict need-to-know basis, one document makes clear that with some decryption, "there will be NO need-to-know."[95]

Not all cyberattacks become public, and the following is not an inclusive list of all cyberattacks conducted by the United States on other states. These are, however, some of the more important instances of cyberattacks.

The earliest known incident of a cyberattack on an industrial control system was the alleged 1982 CIA sabotage of the Soviet Trans-Siberian pipeline infrastructure system. The CIA, working clandestinely with a Canadian supplier, inserted a "logic bomb," which caused the pipeline to explode.[96] In 2017, Russian state-owned international news organization, RT (formerly *Russia Today*), acknowledged that there were numerous American cyberattacks on the Russian president's website, on government servers, and on control systems of energy and telecommunication infrastructure. Moreover, the United States threatened cyber retaliation against Russia in the case of a major cyberattack against the United States.[97]

In 2012, Russian president Vladimir Putin specifically blamed then-US secretary of state Hillary Clinton for a reputed disinformation cyberattack against him. Following the December 2011 parliamentary elections and through the presidential election on March 4, 2012, Russia witnessed domestic protests that were organized largely over social media and blog sites.[98] The protests were held in Moscow's Bolotnaya Square and focused on the fairness of recent elections. "She set the tone for some of our actors in the country and gave the signal," Putin said, referring to Clinton's comments regarding the legitimacy of the December vote. "They heard this and, with the support of the US State Department, began active work."[99] While the specifics of any potential US role in the Bolotnaya Square protests have never been clarified by either the United States or Russia, there was strong evidence of grassroots Russian grievances based on a series of earlier protests.[100] Thus, Putin's accusations leveled against the State Department draw on a well-known pattern of activity, whether or not they carry any weight, is difficult to assess. Although the DoD is heavily involved in public diplomacy, it is the legal responsibility of the US Department of State[101]—the use of interactive internet activities, regionally focused websites, and social media to advance American interests overseas—following the 1998 decision to fold the US Information Agency into the State Department with the end of the Cold War.[102]

The issues surrounding the current challenges of public diplomacy in a cyber era were outlined in a 2009 book by the US Congressional Research Service.[103] *US Public Diplomacy: Background and Current Issues* states:

> Internet communications, including social media networks such as Twitter and Facebook, have characteristics of both broadcast communications, such as the ability to communicate written and spoken words, still images, and motion pictures to a wide audience and in-country, person-to-person outreach, which engenders personal relationships connecting networks of individuals connected by common interests, not just common geography.[104]

This has not always gone as well as it could. For instance, a US broadcaster in the Middle East, Alhurra, was criticized after it "allowed terrorist organizations and Holocaust deniers to promote their views on the air; it is argued, has damaged Alhurra credibility with the Arab public."[105]

Less convincing were Russian concerns about foreign influence on a site called Kartanarusheniy—an interactive map of election violations—sponsored by the Russian independent election watchdog Golos (voice).[106] Kartanarusheniy was taken down in the spring of 2014, as were the sites that linked to it or mentioned it.[107] However, Golos applied to the Russian Justice Ministry demanding to be excluded from the list of foreign agents after the Moscow city court ruled that the association had not violated the law, citing the constitutional court ruling of April 2014, according to which a nongovernmental organization is not obliged to register as a foreign agent if it refuses foreign funding. The Moscow court concluded that the unsuccessful attempt by the Norwegian Helsinki Committee, a nongovernmental organization working to ensure that human rights are respected,[108] to transfer money to Golos "is not sufficient grounds to conclude that the association receives foreign funding, which is one of the requirements for listing an NGO as a foreign agent."[109] The Russian government's attempt to use the specter of Western interference to silence the opposition was obvious even to other Russian actors like the Moscow city court.

The United States is, however, stepping up digital incursions into Russia's electric power system network. Since at least 2012, current and former officials say, the United States has put reconnaissance probes into the control systems of the Russian electric grid. However, now the US strategy has shifted more toward offense, with the placement of potentially crippling malware inside the Russian system at a depth and with an aggressiveness that had never been tried before. It was partly a warning meant to be discovered, because deterrence works if the other side knows that there can be serious retaliation, and partly in preparation to conduct cyberattacks if a major conflict occurred—a demonstration of how the Trump administration is using new authorities to deploy cyberweapons more aggressively.[110]

This covert measure that authorized implanting code in sensitive computer systems that Russia was bound to find to serve as a reminder of America's cyber reach—as well as the expulsion of thirty-five Russian diplomats, the closure of two Russian-owned compounds in Maryland and California, and economic sanctions on Russian intelligence officials—was part of a retaliation package approved by President Obama in late December 2016.[111] The retaliation was in response to an August 2016 report drawing from sourcing deep inside the Russian government that detailed President Putin's direct involvement in a cyber campaign to disrupt and discredit the US presidential race, including Mr. Putin's specific instructions on the operation's objectives to defeat or damage US presidential candidate, Hillary Clinton, and help elect then-US presidential candidate, Donald Trump.[112]

From August to December 2016, the Obama administration's Cyber Response Group debated dozens of options for deterring or punishing Russia.[113] Simultaneously, there were at least five warnings[114] to the Russians not to intervene in the actual election, including a news conference on September 5, 2016, when President Obama issued a veiled threat: "Frankly, we've got more capacity than anybody both offensively and defensively."[115] Finally, on October 31, 2016, the administration delivered a final preelection message via a secure Cold War-era nuclear channel, stating that the United States had detected malicious activity from Russian servers targeting the US electoral systems and warned that meddling would be regarded as unacceptable interference.[116]

The American non-cyber retaliation to Russian cyberattacks during the November 2016 presidential elections included hybrid elements. Although it was not specifically kinetic war, there were physical reactions to the interference, including the detention of four Russian men that the Americans accused of cyberattacks. One is a Russian citizen, Yevgeniy Nikulin, originally held in Prague, in the Czech Republic, on an Interpol arrest warrant issued by US authorities. Although both the United States and Russia requested his extradition,[117] he was extradited to the United States on March 30, 2018, and was later questioned regarding Russian cyberattacks on US elections.[118] A second Russian computer programmer, Stanislav Lisov, was arrested by Spanish police at the Barcelona airport in January 2017 on another US warrant. He was extradited to the United States on January 19, 2018. A third Russian citizen, Roman Seleznev, was extradited to the United States from Guam amid Russian protestations in 2014 and was convicted in 2016 on thirty-eight cyber-related charges by a US court.[119] A fourth Russian programmer, Pyotr

Levashov, was arrested under an international warrant regarding cyberattacks linked to the interference in the 2016 American presidential election.[120] He was extradited to the United States on February 2, 2018.[121] These four cases come in addition to the thirteen individuals and three companies that were indicted as part of the US Special Counsel Robert Mueller's investigation on February 16, 2018.[122] Additionally, the investigation indicted twelve Russian intelligence officers from the GRU on July 13, 2018, who will never be extradited to the United States.[123]

One of the best-known US cyberattacks against adversaries was the attack on Iran's Natanz nuclear facility called Operation Olympic Games, mentioned in an earlier chapter. Less well known are the June 2019 US Cyber Command attacks against multiple computer systems. One target was an Iranian intelligence group believed to be behind attacks against oil tankers, owned by Norway and Japan, in the Gulf of Oman. Another targeted computer systems that control Iranian missile launches. The cyber operation was intended to be below the threshold of armed conflict, using similar shadow tactics to those deployed by Iran. The cyberattacks, which had been planned for several weeks, were ultimately meant to be a direct response to both the tanker attacks and the downing of a US drone in June 2019.[124] Iran claimed the drone was in its airspace, while the United States insisted it was shot down in international airspace.[125]

On April 16, 2017, a North Korean missile test blew up seconds after liftoff under suspicions that a covert US cyber program to sabotage the test flights had succeeded again. The attacks on North Korea's missile program, which the Obama administration hastened in 2014, resulted in 88 percent missile failures.[126] It can be difficult to determine if an individual launch is the result of a cyberattack, even inside the US Cyber Command and the NSA, where the operation is centered. However, evidence suggests that North Korea, using a different kind of missile, has overcome at least some of the problems.[127]

The clandestine US cyber operation targeting the North Korean missiles was enthusiastically adopted by the Trump administration to the point of openly discussing it with the president of the Philippines, Rodrigo Duterte.[128] In addition to the missile component of the cyber campaign, US Cyber Command targeted North Korea's military spy agency, the Reconnaissance General Bureau, by barraging its computer servers with a DDoS attack that choked off internet access. The effects were temporary and not destructive.[129]

To focus on cyberterrorism, US Cyber Command has launched cyberattacks on the so-called Islamic State (ISIS) group. The mission was less than successful. The cyberattack attempted to disrupt communications, recruitment, payroll, and directives. However, new recruitment efforts and communication hubs reappeared almost immediately. The most sophisticated US offensive cyberattack against the Islamic State beginning in November 2016, Glowing Symphony, sabotaged the group's online videos and propaganda, but they too were quickly replaced. Cyber tactics must shift to more effective targeting of financial assets and to compromise leadership in order to effectively fight terrorists who are very adept at social media, careful with backing up files, and firmly in control of their own online media presence.[130]

Nonetheless, some effective cyberattacks, by the newly created cyber mission teams that joined traditional military units in hybrid warfare, kept ISIS fighters and commanders from seeing or having advanced notice of traditional physical attacks.[131] One of the rare successes against the Islamic State was by Israeli cyber operatives who penetrated a small cell of bomb makers in Syria. They exposed that the terrorist group was making explosives that looked like laptop computer batteries to fool airport screening. The intelligence prompted a ban on large electronic devices in carry-on luggage on flights in March 2017. It was this classified intelligence that President Trump revealed when he met in the Oval Office with the Russian foreign minister, Sergey Lavrov, and the ambassador to the United States, Sergey Kislyak, in May 2017. This disclosure of classified information infuriated Israeli officials.[132]

When cyberespionage and cyberattacks merge to create a back door as a precursor to attack, this is known as "exploitation" under the US cyber strategy.[133] A classic example of this merging into exploitation is the protection of the US electrical grid and other infrastructure. The US grid has three electric networks—one in the East, one in the West, and one in Texas—with thousands of miles of transmission lines, power plants, and substations. The flow of power is controlled by local utilities or regional transmission organizations. As the electrical grid and other utilities rely on online communication, these utility control systems are vulnerable to cyberespionage, cyberattacks, and exploitation.[134] Under the George W. Bush administration, Congress approved $17 billion for the protection of government networks. Under President Obama, the program continued to pour billions of dollars into addressing vulnerabilities in private computer networks. Cyberespionage probes—from China, Russia, and elsewhere—have navigated the US electrical system without damaging the power grid but in preparation for cyberattacks.[135] Pre-positioning malicious software in critical systems includes the HAVEX[136] and BlackEnergy[137] malware placed by Russia, discovered in the US electrical grid in 2013.[138] This blurred line between cyberespionage and cyberattacks is characteristic of shadow warfare and is utilized by all adversaries against each other with relative success.

CONTROLLING CYBERSPACE

The ability to control cyberspace was announced in the "International Strategy for Cyberspace: Prosperity, Security, and Openness in a Networked World," published by the White House in 2012, which stated that "the digital world is no longer a lawless frontier."[139] The Wild West quality of cyberspace was now a thing of the past as far as the United States was concerned. The press started reporting on the US plan to control cyberspace in 2012, specifically in regard to DARPA.[140] DARPA's Plan X has an advanced map that details the entirety of cyberspace—a global domain that includes tens of billions of computers, networks, cyberweapons, and bots—and updates itself continuously. The ideal map shows network connections, analyzes how much capacity a route has for carrying a cyberweapon, suggests alternative routes according to traffic flows, and indicates power and transportation systems that support military objectives. Plan X allows a visual representation of cyberspace to assist decisions on what to attack and how, while seeing any attacks coming from an adversary. Plan X hardens operating systems capable of launching attacks and withstanding retaliation.[141]

Control over cyberspace has led to a greater ability to assign cyber operations to specific states and specific operators. Thus, cyber operations lose much of their secret non-attributable qualities. Every kind of cyber operation—malicious or not—leaves a trail. US intelligence analysts use their constantly growing knowledge of previous events, of how cyber operators work, and of existing and emerging cyber tools to trace cyber operations back to the point of origin and often to specifically named operators.[142]

A central aspect of regulating cyberspace is domestic control. The United States is largely a regulatory state rather than a state that controls from the top down. The primary regulation is the Federal Communications Commission (FCC) authority through a 1996 law that represents major legislation on communications policy. The landmark Telecommunications Act of 1996 was passed by a bipartisan majority, the Clinton administration, and the FCC. The act left US governance of cyberspace to the engineering-driven multistakeholder process that created it— and free of political management.[143] In 2015, the Obama administration enacted rules prohibiting internet providers from charging more for certain content or from giving preferential treatment to certain websites. In 2018, the FCC repealed the 2015 net neutrality rules.[144] This leaves much of cyberspace control in the United States to commercial entities.

Economic aspects within cyberspace have become extremely important in the United States. The Telecommunications Act states that "it is the policy of the United States to preserve the vibrant and competitive free market that presently exists for the Internet and other interactive computer services, unfettered by Federal or State regulation."[145] With the Trump administration's 2018 repeal of the 2015 law, e-commerce start-up companies feared that they could end up on the losing end of paid prioritization, with their websites and services loading more slowly or with a lower priority than those run by internet giants. Remote workers, including freelancers in the gig economy, could similarly face higher costs to do their jobs from home. The FCC said it had repealed the rules because they restrained giant broadband providers from experimenting with new business models and investing in new technology.[146]

Controlling cyberspace similarly focuses on military operations in and through cyberspace. The US military recognizes that most states maintain that their national cyberspace is considered a sovereign domain, but this is not necessarily restrictive. According to the US military's doctrine outlined in the 2018 *Cyberspace Operations*, "In cyberspace, there is no stateless maneuver space. Therefore, when US military forces maneuver in foreign cyberspace, mission and policy requirements may require they maneuver clandestinely without the knowledge of the state where the infrastructure is located."[147] According to this document, the military conducts cyber operations "consistent with US domestic law, applicable international law, and relevant [governmental] and [military] policies. The laws that regulate military actions in US territory also apply to cyberspace."[148]

Perhaps the most pertinent perspective regarding controlling cyberspace is Washington's international stance. In this regard, the United States is in one of the two main camps divided over how international cyber operations should be organized and legislated. On the one side is the Western camp, which focuses on applying existing international law to cyber operations and, on the other is the camp that calls for creating specific international laws and treaties and reinforcing international political structures as the mechanisms for maintaining international peace and security in the era of shadow warfare.[149]

The United States has four arguments for rejecting the need for any new treaty for maintaining international peace and security regarding international cyber operations. First, the United States argues that it would be "premature to formulate overarching principles pertaining to information security in all its aspects,"[150] meaning that, in a rapidly developing cyber technology era, it is too soon to even consider such a treaty. The second argument is partly that a multilateral treaty for restricting the development or use of cyber civil and/or military technologies was unnecessary, as the law of armed conflict—especially the principles of necessity and proportionality—was already applicable to cyberweapons and cyberwarfare. The third argument is partly that cyber technologies not directly tied to warfare were best addressed in international committees better suited for discussion of subjects other than disarmament and international security. Finally, the United States argues that a treaty approach was against the principle of the free flow of information critical to the growth and development of all states: "The implementation of information security must not impinge upon the freedom of any individual to seek, receive and impart information and ideas through any media—including electronic—and regardless of frontiers, as set forth in Article 19 of the Universal Declaration of Human Rights."[151]

There is another aspect to the US argument against the need for a new international treaty on cyber operations. In addition to the abovementioned justifications, the United States asserts that its view on self-defense, including preemptive self-defense, applies to cyberwarfare. The US view is that nothing impairs the inherent right of individual or collective self-defense if an armed attack occurs; this includes the right to respond to a particularly devastating cyberattack with a kinetic retaliatory counterstrike. The United States has adopted a broad concept of self-defense that justifies preemptive military action. Indeed, the Trump administration's April 2017 strike on a Syrian government airbase in retaliation for a chemical weapons attack that killed dozens of civilians in a rebel-held town confirms that the US government does not feel itself limited to a strict understanding of self-defense. Within this understanding, it is not unreasonable to imagine a US administration justifying a cyberattack on preemptive self-defense grounds.[152]

Moreover, the United States and other members of this camp assert that international humanitarian law is applicable to cyberattacks as well, especially in case a threat of force or use of force occurs in an incident that causes humanitarian suffering. For instance, if a cyberattack takes down a country's or an ally's electric grid, like what happened in the Ukrainian capital Kyiv in December 2015, international humanitarian law can justify either a counter-cyberattack or even a kinetic response to a cyberattack that could escalate as far as a potential[153] nuclear response.[154] As David Sanger details in *The New York Times*, one extreme example might include a US nuclear strategy in the Trump administration that would permit the use of nuclear weapons to respond to crippling cyberattacks on US infrastructure.[155] The response, of course, would have to be proportional, so the level of retaliation—especially for a nuclear response—would imply an attack that was nationally debilitating.

US confidence that it can control cyberspace, both domestically and internationally, certainly underlies this stance. The United States does not close off its domestic cyberspace, primarily for reasons of the freedom of flow of information. Rather, the United States relies on retaliatory responses to cyberattacks either through the existing laws of self-defense or by the use of international humanitarian law. This reflects its stance that no international treaty which would restrict these retaliations is needed or even desirable. Instead, the United States relies on its own sophisticated understanding of cyberspace and its own ability to construct retaliatory weaponry.

SHADOW WARFARE POLICY

When it comes to developing US shadow warfare policy, three central components always become pertinent. First and foremost are the political, economic, and social factors that influence long-term goals that guide policy, both implicitly and explicitly. Second, policy is largely a product of individual administrations, each having its own unique character. Finally, principles, both national and ideological, also shape policy.

Fundamentally, US shadow warfare policy is guided by political, economic, and social realities. Politically, the United States is still a leader in most warfare technology, including cyberwarfare. This disinclines the United States from voluntarily constructing limits on its technological capabilities. In particular, the United States argues that a treaty along the lines of those negotiated for chemical or nuclear weapons is unnecessary. Instead, US officials make a negative argument that there should be international law enforcement efforts to confront the rise in cyber violence, especially in criminal cyber activities—including cybercrime, hacktivism, cyberterrorism, and cyber disinformation campaigns—through improved cooperation. The idea is that this covers many of the cyberattacks against the United States and addresses some of the defensive actions that fall short of kinetic war retribution or even kinetic war targets.

US officials also make a positive two-pronged argument that a traditional cyber treaty is unnecessary. The first argument is in support of openness and interoperability.[156] The *Prosperity, Security, and Openness in a Networked World* strategy states its reliance on an earlier 2005 agreement, saying that

> the United States supports an Internet with end-to-end interoperability, which allows people worldwide to connect to knowledge, ideas, and one another through technology that meets their needs. The free flow of information depends on interoperability—a principle affirmed by 174 nations in the Tunis Commitment of the World Summit on the Information Society. The alternative to global openness and interoperability is a fragmented Internet, where large swaths of the world's population would be denied access to sophisticated applications and rich content because of a few nations' political interests.[157]

The second positive argument is that a traditional cyber treaty undermines the role of norms and shared understandings in shaping acceptable behavior. Again, the document continues:

> Adherence to such norms brings predictability to state conduct, helping prevent the misunderstandings that could lead to conflict. The development of norms for state conduct in cyberspace does not require a reinvention of customary

> international law, nor does it render existing international norms obsolete. Long-standing international norms guiding state behavior—in times of peace and conflict—also apply in cyberspace. Nonetheless, unique attributes of networked technology require additional work to clarify how these norms apply and what additional understandings might be necessary to supplement them.[158]

Second, US policy is made primarily by the administrations within the executive branch, rather than resting in the legislative or judicial branches. The legislative branch would be involved if there was a declaration of war, but shadow warfare eschews declaration almost by definition. Therefore, the oversight is within the various departments within the executive branch by department secretaries, military leaders, and sometimes by the White House itself. This means that often oversight is minimal and often unseen by lawmakers and citizens to the extent that it does exist.

In an ongoing process to create policy for shadow warfare and its principal tool of cyberwarfare, the United States continues to rely upon Carl von Clausewitz's *On War*. Von Clausewitz's ideas are relevant to cyberwarfare and to the combination of cyberwar with traditional kinetic war to create hybrid warfare. Clausewitz writes in his first chapter: "War is an act of violence, which in its application knows no bounds; as one dictates the law to the other, there arises a sort of reciprocal action, which, in the conception, must lead to an extreme."[159] According to René Girard in *On War and Apocalypse*, Clausewitz put a finger on an irrational aspect of reality—that is, the world tends to extremes in war.[160] These extremes, once thought to have reached the limit first in total war and then in nuclear war, now usher in the continuous shadow of cyberwarfare.

Just war theory is interpreted to support the defining weapon of shadow warfare—cyberwarfare. Although some scholars, such as Randall Dipert in *The Ethics of Cyberwarfare*,[161] argue that just war theory does not straightforwardly apply to cyberwarfare when cyberwar does not kill humans, Colonel James Cook, a professor at the US Air Force Academy, persuasively asserts that analogously ambiguous cases have long existed in warfare without undercutting the just war theory's broad relevance.[162] The characteristics of cyberwarfare fit the confines of just war. The norms of just war—with *jus ad bellum* criteria, including just cause, comparative justice, legitimate authority, last resort, and probability of success—blend with *jus in bello*, including right intentions, proportionality, and noncombatant immunity. Consider the principle of defense in the face of aggression. Cyberattacks are aggressive when it comes to preventing a state from meeting basic human needs, like compromising a state's ability to provide electricity as a result of malware in power grids. The second principle of just war is the protection of noncombatants. Effective cyberattacks search for targets and spread the attack but, as with biological warfare, have a reasonable probability of spreading to noncombatants, such as in the case of Stuxnet. The third principle of just war is proportionality—the idea that it is wrong to cause more harm in defending against an attack than the harm caused by the attack itself, which was reflected in the US response to Russian meddling in the US election. US cyberwar theory is a progression from the early Western evolution of the just war theory and new interpretations of von Clausewitz for a cyber age.

In sum, the United States regards the rules surrounding just war theory and Carl von Clausewitz's concepts of war to be the foundation for shadow warfare theory and policy and its primary tool of cyberwarfare. The United States rejects the current need for a cyber treaty, arguing negatively that the concern should be on cyber violence, which is best addressed by harmonizing cybercrime laws internationally and by enhancing international cooperation in cybercrime.[163] The United States also argues that treaties limit offense at a time when the United States is concerned with cyber defense. Positively, the United States sees cyberspace as having a constructive impact on the global quality of life and considers the existing norms of international law and just war theory as sufficiently applicable to cyber operations, thus limiting the need for an international cyber treaty. The fact that Russia and China have vastly different notions of shadow warfare theory does not factor into current US thinking on cyberwarfare and the need for an international agreement.

Chapter 3

Cyber Russia

As shadow warfare becomes increasingly relevant in power projection, national security, and strategic planning, it is not surprising that the three big players in the field are the three big players in many aspects of security: the United States, China, and Russia. This chapter looks at the national policies and the underlying doctrines of the Russian Federation, as well as examples of cyberattacks, such as who is being attacked, how, and why. Russia plays a big role in global shadow warfare, with a specific set of policy preferences and a distinctive worldview.[1]

As discussed in the previous chapter, with the ongoing state of perpetual shadow warfare, the rivalry of the great powers is playing out with hints of earlier rivalries. In some respects, much as cyberwarfare's lack of international norms and rapidly developing national policies, there is a return to elements of Cold War rivalry that has morphed into this new type of clandestine conflict. While the Cold War had deeply entrenched ideological conflicts, the concept of allies and adversaries is much more fluid in shadow warfare. Moreover, the concept that this is war without a clear end is similar to the Cold War mentality; the reality is that cyberwar, like nuclear war, will not end. It is interesting to consider that the notion of perpetual war does not necessarily presume a perpetual enemy, as Jānis Bērziņš, director of the Center for Security and Strategic Research at the National Defense Academy of Latvia and one of the leading specialists on Russian military strategy in the world, suggests.[2] It does, however, fundamentally change the notion of who and what makes an ally or a foe. For instance, the extraordinary variations allowed in hybrid war, which has been well developed by Russia, mean that some allies are also partial adversaries, and adversaries can occasionally be partial allies in the swiftly shifting sands of shadow warfare.

One of the issues with shadow warfare is that new powers and new weapons present the need for new strategies, new institutions, and new ways of looking at allies and adversaries as well as new policy challenges. Russia is adapting its own national considerations to a new regime, with some consistent factors from earlier traditions. The approach has been unique to Russia, but with some elements that reflect the increasingly common standards of shadow warfare and its primary weapon of choice—cyberattacks.

Outlining the use of cyber instruments to protect itself and its national interests, the Russian government released its military doctrine in February 2010. The document defines a modern military conflict as including the integrated use of military and nonmilitary capabilities and a greater role for cyberwarfare, which Russia calls information warfare. Specifically, the doctrine summarizes a belief that cyberwarfare can achieve political objectives without using military force and spread information and disinformation that sways the international community to act in ways that benefit Russia. This understanding of the uses of cyberwarfare means the creation of new forms of offensive cyberweapons and the creation of new forms of defensive cybersecurity.[3]

In 1997, the Russian Federal Agency of Government Communications and Information outlined cyberwarfare as having four components: the destruction of command and control centers combined with an electromagnetic attack on information and telecommunications systems; the acquisition of intelligence; disruption of computer systems; and disinformation.[4] The 2000 Information Security Doctrine of the Russian Federation was readopted in 2008 and remained in force until December 2016, when a new Doctrine on Information Security of the Russian Federation, with a slightly different title, was adopted.[5] The basic elements remain, but the concepts of cyberwar have been refined. The three pivotal objectives outlined in the doctrine are establishing full state control over the domestic cyberspace, overcoming the international "discrimination" of the Russian media, and growing concerns that Russia is lagging behind other key players in the domain of information technology and cybersecurity.[6]

President Vladimir Putin has personally set up Russia's cyber institutions and strategies and has personally directed many of the cyberattacks beginning in his first two terms in office between 2000 and 2008, and then with a much more sophisticated systematic gearing up of cyberattacks and other shadow warfare strategies from his second round as president, beginning in 2012 and still ongoing.[7] According to the special incident report drawn up by the German intelligence services, major attacks such as those directed against the German Bundestag and the US presidential election in 2016 were likely "directly authorized by the presidential administration in the Kremlin and left up to the services to carry out."[8] As his 2012 "Russia and the Changing World" speech illustrates, Putin remains focused on information and disinformation:

> World public opinion is being shaped by the most active use of advanced information and communications technology. . . . This implies a matrix of tools and methods to reach foreign policy goals without the use of arms but by exerting information and other levers of influence. Regrettably, these methods are being used all too frequently to develop and provoke extremist, separatist and nationalistic attitudes, to manipulate the public and to conduct direct interference in the domestic policy of sovereign countries.[9]

With this speech, Putin set the basis of Russia's cyber institutions and cyber strategy.

INSTITUTIONS AND INDIVIDUALS

This institutionalization of shadow warfare occurred in the aftermath of the collapse of the Soviet Union in 1991 as the new Russia was taking shape. The world was changing, and new technologies, including cyberspace, were coming into being. In this new era, Europe and the United States changed their policies of containment and isolation toward Russia to an approach of economic and political integration. Russia began to rely on trade with Europe, which in turn began to rely on Russian energy. By 2000—and the start of Putin's reign—Russian economic integration shifted from a positive force to a corrosive negative one. After the 2008 global financial crisis, the United States and the European economies relied increasingly on Russian fossil-fuel wealth for funding and investment. At the same time, Russia directed its institutions to enhance its impact on American and European political and economic systems in order to advance its national interests through shadow warfare.[10]

This institutionalization of Russian cyber power, however, is not absolute. While Russian government authorities give specific and detailed orders and instructions regarding shadow warfare attacks, there appears to be rivalry and competition within the system—including in the intelligence services. Moreover, execution is often loosely organized and delegated to a broad variety of actors. Some are tied closely into a chain of command, and others are linked much more tenuously to government authorities—subcontractors, businessmen, privateers, and even organized cybercrime networks—allowing for maximum agility, speed, adaptability, and creativity. It permits proceeding by trial and error, and it allows state actors to evade attribution—and retaliation. The example of the use of Russian organized crime, while not exclusive to Russia, has many unique characteristics. In addition to the security agencies' cyber abilities, Russia still depends, to a considerable extent, on recruiting cybercriminals or simply calling on them from time to time. There have been several other attacks that look like the work of organized crime, such as the 2010 infiltration of the NASDAQ's central systems.[11] Organized crime also provides surge capacity for cyberattacks, such as the DDoS attacks on Estonia in 2007 and Georgia in 2008 and ongoing cyber disruption in Ukraine.[12] As the opportunities for cyberattacks and cyberespionage grow, governmental cyber forces are continuing to outsource organized crime in Russia.[13] However, it can also mean a sacrifice of control and effectiveness,[14] such as the 2017 WannaCry cyberattack, where the Russian government was unable to protect itself, and thousands of Russia's corporate and government networks were hit.[15]

Russian cyber capabilities, like those of the other major cyber powers, are evolving and adapting via a spectrum of institutions.[16] Russian cyber institutions were initially characterized by a unique interconnection of government, business, and crime.[17] However, the initial DDoS cyberattacks that once typified Russian cyber operations have been supplemented by more sophisticated tactics and malware tools. Russia maintains strong partnerships with industries and universities to assist with the research and development of cyber capabilities.[18]

Playing a more direct role in offensive cyber operations than in the past are two state institutions: the Main Intelligence Agency of the General Staff of the Russian Armed Forces (GRU), Russia's military intelligence, and the Federal Security Service of the Russian Federation (FSB), the principal security agency and successor to the KGB.[19] The FSB absorbed parts of the Federal Agency of Government Communications and Information, which had been responsible for cryptology and code-breaking.[20] The GRU has clear links to advanced persistent threat groups APT28/Fancy Bear and APT29/Cozy Bear, and to Russian cyberattack and propaganda groups.[21] The GRU's APT28/Fancy Bear includes Units 26165 and 74455, which engaged in cyber operations that involved the staged releases of documents stolen through computer intrusions along with APT29/Cozy Bear, including interference with the 2016 US presidential election and the 2018 midterm elections.[22] The same unit has been involved in attacks on French president Emmanuel Macron, the North Atlantic Treaty Organization (NATO), the German parliament, Georgia, and other government targets across Europe.[23]

Interestingly, the Dutch General Intelligence and Security Service (AIVD) penetrated the APT29/Cozy Bear computer servers and a security camera at the entrance of their offices in a university building in Moscow. The Dutch were the first to alert their US counterparts about the Russian cyberattacks gaining access to US agencies and the Democratic National Committee

(DNC). This account is based on interviews with a dozen Netherlands and US political, diplomatic, and intelligence sources with direct knowledge of the matter.[24]

These activities were the result of a shift in cyber operations after the creation of a Cyber Command in the beginning of 2012. Russia's Cyber Command was conceived as "a super special purpose force" that was supposed to be eventually expanded to tackle a wide range of tasks. The creation of the Cyber Command was part of the most ambitious, consistent, and effective military reform, called the New Look, begun in 2008. The crisis in Ukraine became the first significant test for this new modus operandi. The covert Russian military intervention in Crimea in 2014 had some of the observers in the West and Ukraine talking about a new hybrid war.[25] While Russia had resorted to hybrid war several times in the past, current hybrid warfare—such as economic manipulation, an extensive and powerful disinformation and propaganda campaign, the fostering of civil disobedience and even insurrection, and the use of well-supplied paramilitaries—supports Russia's conventional forces.[26]

Russia centered offensive and defensive cyber capabilities in the military by establishing special military cyber units and a cyber coordination unit—the Cyber Defense Center—subordinate to the GRU.[27] In 2013, a senior Russian official in the research and development wing of the military confirmed that Russia was enhancing its cybersecurity and creating a separate cyberwarfare wing under the jurisdiction of the military, which sees cyberspace as the new theater of war.[28]

Like the other major players, Russia is institutionalizing cyber warfare and privateers. Russian state agencies are clearly conducting advanced network reconnaissance and developing malware to attack specific system vulnerabilities as other governments harden their networks. For instance, the cyberattacks perpetrated ahead of the outbreak of conflict in Georgia and Ukraine were facilitated by spear-phishing cyberespionage campaigns that introduced malware or granted cyber actors' remote access to systems in anticipation of a potential future military or diplomatic action.[29]

One notorious group of privateers has gone under several names but is best known as the Internet Research Agency (IRA). The IRA is an organization in St. Petersburg, Russia, that spreads disinformation on the internet. It employs hundreds of Russians as paid social media users to post pro-Russian government propaganda online under false identities—including buying intentionally divisive ads and extensive identity theft. The IRA, along with several of its employees, was indicted by the FBI for committing federal crimes while seeking to interfere in the US political system, including the 2016 presidential election.[30] According to the FBI[31] and several Russian media outlets, the IRA is funded by Evgeny Prigozhin, an oligarch restaurateur known as the

"Kremlin's chef" due to his large government contracts and his close relationship with President Putin. Prigozhin's Concord holding company was shown to fund the IRA when illegally accessed emails revealed an accountant at Concord approving payments to the agency.[32] Leaked documents listed Mikhail Burchik, a retired St. Petersburg police colonel who joined the agency in February 2014, as its executive director,[33] indicating another level at which it is connected to the government.

CYBER STRATEGY

Russia takes a broad approach to cyberwarfare strategy or information operations. Cyberwarfare strategy in Russia includes the cyberespionage aspects of intelligence gathering and counterintelligence; the cyberattack elements of electronic warfare, debilitation of communications, degradation of navigation support, and degradation of information systems; and social cyberattacks that include disinformation, psychological pressure, and propaganda. In Russian information warfare, computers, privateers, and bots are paired with news outlets, social media, and hacktivist communities.[34] DDoS attacks, social media bots spreading disinformation, and the RT television channel (formerly Russia Today) propagating Russian government views are all interconnected tools of cyberwarfare,[35] with Russian news services focusing on the same issues as the disinformation campaigns, such as race in America and the relationship with the European Union in Britain.

The Russian strategic doctrine, New Generation Warfare, is "primarily a strategy of influence," and its primary goal is "breaking the internal coherence of the enemy system—and not about its integral annihilation."[36] The New Generation Warfare is simple and straightforward: it penetrates and utilizes the system from within, for example, preventing the work of an independent press and judiciary and allowing media outlets to disseminate erroneous disinformation that fosters public confusion and disillusionment.[37] No one knows what to believe anymore.

Moreover, New Generation Warfare concentrates "on the exploitation of state resources to further Russian networks of influence."[38] This becomes crucial when realizing that Russia's primary strategy involves "actively discrediting the Western liberal democratic system as well as offering the alternative"[39] of its own highly centralized system of the political and economic power structure. The strategy aims to undermine the cohesiveness and stability of the Western world order.[40] This is a grand strategy, and it may well be working, given the intense polarization in the United States around the result of the 2016 presidential election and round Brexit in the UK.

Hybrid war is also central to Russian strategy. Russia may have been the first of the three major players to fully integrate modern cyberattacks into hybrid warfare, albeit the United States had been doing this in a small way at the start of the twenty-first century. A strategy that paired cyberattacks and cyber disinformation operations with kinetic, physical military attacks—a true hybrid warfare strategy—was launched by Russia in about 2007 and has since been refined and expanded.[41] Then, in 2008, cyberattacks were paired with kinetic attacks in Georgia in a sophisticated orchestration of the cyberattacks with physical attacks. The use of hybrid warfare, a mix of combat, intelligence, and propaganda tools, was deployed in conflicts such as Syria, with the Syrian Civil War cited as an example of successful Russian intervention abroad.[42]

Cyberwarfare is also central in the Russian strategic deterrence framework. The Russians are penetrating industrial control networks responsible for operating critical infrastructure in opponent countries.[43] The strategic objective in the deterrence framework is to develop the capability to remotely access and disrupt the control systems of adversaries in the event of increased hostilities, whether those hostilities are cyber, kinetic, or hybrid.[44]

The theoretical and doctrinal underpinnings of the Russian approach to cyberwarfare reflect the conviction that Russia is locked in a constant struggle with internal and external adversaries seeking to challenge its security in the information realm. Its cyberwarfare doctrine ignores distinctions between peacetime and wartime. This suggests that Russia has a relatively low bar for employing cyberweapons and disinformation in ways that might be viewed as threatening and escalatory in nature.[45] The shadow warfare concept of low-grade, continual warfare is well established in the Russian strategy.

New Generation Warfare, which operationalizes Russian military strategy, outlines a strategy that wins hearts and minds first, with kinetic operations following. The main guidelines for Russian military capabilities by 2020 are the following:

> 8 i. From direct destruction to direct influence; ii. from direct annihilation of the opponent to its inner decay; iii. from a war with weapons and technology to a culture war; iv. from a war with conventional forces to specially prepared forces and commercial irregular groupings; v. from the traditional (3D) battleground to information/psychological warfare and war of perceptions; vi. from direct clash to contactless war; vii. from a superficial and compartmented war to a total war, including the enemy's internal side and base; viii. from war in the physical environment to a war in the human consciousness and in cyberspace; ix. from symmetric to asymmetric warfare by a combination of political, economic, information, technological, and ecological campaigns; x. from war in a

defined period of time to a state of permanent war as the natural condition in national life.[46]

Thus, the Russian view of shadow warfare is based on the idea that the main struggle is for the mind. As a result, new-generation wars are dominated by disinformation and psychological warfare. In order to achieve superiority on the ground and take control of weapons systems, the Russian government wants to morally and psychologically depress an adversary's military personnel and civilian population. This has strong historical parallels to Russian use of colorful depictions in posters and press stories in World War I to raise morale at home, regardless of the realities at the front, and to demoralize the enemy by demeaning ridicule.[47] The main objective is to minimize the necessity for deploying kinetic military power, making the opponent's military and civil populations support the attacker to the detriment of their own government and state.

The New Generation Warfare doctrine combines subtle and discrete state involvement with explicit direct involvement. Aspects of the strategy were evident earlier but were practiced and refined in Ukraine. As outlined by Phillip Karber and Joshua Thibeault in their article, "Russia's New-Generation Warfare," the strategy includes five elements. First, New Generation Warfare uses political subversion such as insertion of agents, political propaganda, and modern mass media to exploit existing ethnic, linguistic, and class differences and to play up corruption or compromise local officials. Second, it uses proxy sanctuary, like seizing local governmental centers, police stations, airports, and military depots; arming and training insurgents; creating checkpoints and destroying ingress transportation infrastructure; cyberattacks compromising communications; as well as conducting phony referendums with single-party representation and establishment of a government under Russian tutelage. Third, New Generation Warfare uses intervention, such as the deployment of Russian forces to the border with sudden large-scale exercises involving air, ground, naval, and airborne troops; surreptitious introduction of heavy weapons to insurgents; creation of training and logistics camps close to the border; commitment of so-called volunteer combined-arms tactical groups; integration of proxy troops into higher-level formations that are equipped, supported, and led by Russians. Fourth, the doctrine uses coercive deterrence—secret strategic force alerts and "snap checks"; forward deployment of tactical nuclear delivery systems; theater and intercontinental maneuvers; and aggressive air patrolling of neighboring areas to inhibit their involvement. Fifth, and finally, New Generation Warfare uses negotiated manipulation, which means using Western-negotiated ceasefires to rearm Russia's proxies; using violations to drain an adversary's military capability while inhibiting other states from helping for fear of escalation;

dividing Western alliances by playing on economic incentives; and selective and repetitive phone negotiations with a favorite security partner.[48]

Russian cyber strategy incorporates many aspects of earlier Soviet control over what people think. This has been effective in the past and perfectly suited to cyberwarfare. Soviet control was occasionally done with incredible talent, such as Mikhail Sholokhov's epic *And Quiet Flows the Don*, winner of the 1965 Nobel Prize in Literature, that convinced many Soviet citizens that institutions like collective farms were heroic and reflective of the country's strength and perseverance. It is also reflective of what President Vladimir Putin knows best from his background in the FSB, and its predecessor, the KGB. While this strategy does not avoid pre-positioning malware in adversaries' electric grids, it does focus on the longer-term aims of influencing one's opponents.

CYBERESPIONAGE

Cyberespionage by Russia has been practiced for multiple decades, similar to the United States and China. Under President Vladimir Putin, the Russian government enhanced a domestic system of cyberespionage to influence domestic expression online. It uses media censorship, conducts cyber surveillance, ensures favorable media coverage, intimidates the opposition, and limits discussion of certain topics. The Russian government categorizes surveillance, domestic media policy, cybersecurity, and internet governance as related issues. President Putin set up strategic infrastructures—bots, privateers, and propagandists—to control the domestic message and to intervene in global media systems. Official media outlets such as RT and Sputnik are key parts of that infrastructure, reporting some accurate news combined with disinformation in order to lend credibility to the disinformation.[49]

Domestic cyberespionage in Russia started with the rise of the internet at the end of the twentieth century. The System of Operative-Investigative Measures (SORM) was applied in 1995, requiring telecommunications operators to install hardware provided by the FSB to monitor domestic communications metadata and content, including phone calls, email traffic, and web browsing activity. In 2012, SORM was expanded to include social media platforms. Assumptions are that any information shared on Russian social networks such as VKontakte (in touch) and Odnoklassniki (classmates) is collected by the intelligence services. At least nineteen VKontakte users (76 percent of known cases)[50] were imprisoned in 2018 for posting memes or even liking postings on the site. In 2016, all internet service providers were legally required to install new hardware, essentially giving the government instantaneous access to all information streaming on the Russian internet, or Runet as it is sometimes called. Access to internet information extends to even collecting data from the dating site, Tinder.[51] Data retention and data localization laws collect additional information for SORM. Internet companies must keep communication metadata and content and store all information about Russian citizens on servers physically located within Russia.[52]

Russian domestic cyberespionage appears to have intensified following the anti-government protests of 2011, when tens of thousands of Russian citizens took to the streets after concerns of fraud in the parliamentary elections. The protests were organized largely on social media platforms such as Facebook and Twitter, which was possible because over three-quarters of the Russian population access the internet, primarily via cell phones. While the internet was the one medium where the opposition could get its message out since it did not have access to the mainstream broadcasters and newspapers, this ended after the domestic unrest. In the spring of 2011, DDoS attacks were directed against websites generally associated with opposition to Putin's government. Among the targets were the LiveJournal blog site, websites run by anti-corruption crusader Aleksey Navalny, and the *Novaya Gazeta* newspaper.[53] Laws were passed in 2014 requiring bloggers with more than 3,000 daily readers to register with the media regulator, Roskomnadzor, and internet companies are required to allow Russian authorities access to bloggers' information, which must be stored on servers based on Russian territory for governmental access.[54]

The government can censor websites without a court order as a result of amendments introduced in 2012–2013.[55] Internet platforms like Yandex, a multinational corporation specializing in internet-related services, were subjected to political pressure, while others, like VKontakte, were brought under the control of government allies. President Putin called the internet a "CIA project," one that Russia needed to be protected from. Restrictions online were paired with a new wave of digital propaganda by paid social media users and bots.[56] Presumably, the Russian operations directed at a domestic audience are also designed to show that Europe and the United States are no substitute for Russia and President Putin—that is, the West offers no reasonable alternatives.[57]

A sovereign internet law establishing new controls on the internet came into force in November 2019, giving government officials wide-ranging powers to restrict traffic on the Russian web. While the Putin government said the law will improve cybersecurity, critics fear it is another step along the path to create an internet firewall similar to that of China. The law allows the possibility to switch off connections within Russia or to the World Wide Web. The government decides when connections will be interrupted and require internet service providers to install network equipment capable of identifying the source of traffic and filter content to more effectively control domestic and international cyber realms.[58]

Russia's international cyberespionage was boosted by a family of unique malware toolsets used to steal information by infiltrating computer networks and retrieving stolen data since at least 2008. Targets of the international cyberespionage included the Georgian Ministry of Defense, the Ministries of Foreign Affairs in both Turkey and Uganda, as well as other government institutions and political think tanks in the United States, Europe, and Central Asia. These cyberespionage attacks are the product of a single, large, well-resourced organization that provides the Russian government with intelligence on foreign and security policy matters.[59]

Russian cyberespionage targeted government institutions including embassies, nuclear research centers, and oil and gas institutes. An early Russian cyberespionage initiative, dubbed Red October, gathered sensitive documents from organizations, which included geopolitical intelligence, credentials to access classified computer systems, and data from personal mobile devices and network equipment. It was designed to steal very specific encrypted files. The primary focus of this campaign targets countries in Europe and the former Soviet republics, although there are also North American targets. Like Flame, Red October is made up of several distinct modules, each with a set objective or function. Some modules were designed to target files encrypted using a system known as Cryptofiler—an encryption standard that was once in widespread use by intelligence agencies. While Cryptofiler is no longer used for extremely sensitive documents, it is still used by NATO for protecting privacy and other information that could be valuable to cyber operators.[60]

At the end of the twentieth century, an early Russian cyberespionage operation, Moonlight Maze, systematically broke into the US Department of Defense computers for more than a year and extracted vast amounts of sensitive information.[61] In another operation, Russian intelligence is alleged to use the cybersecurity firm and a popular antivirus provider, Kaspersky, to steal classified information. There are indications that the alleged espionage is related to a public campaign of highly damaging NSA leaks by a stealth group called the Shadow Brokers.[62] A cache of NSA cyberweapons were put up for sale by the Shadow Brokers in 2016, with either Russia as the most likely perpetrator, or an insider's leak, or both. Three NSA employees or contractors have been arrested since 2015 for removing classified files, but there is fear that one or more leakers may still be in place. Russia is the prime suspect in a parallel hemorrhage of cyberweapons and secret documents from the CIA's Center for Cyber Intelligence, posted to WikiLeaks.[63] Edward Snowden tweeted on August 16, 2016, that "conventional wisdom indicates Russian responsibility"[64] for the auction of stolen NSA malicious software files. Snowden's comments imply the auction may be a signal to the

United States to consider carefully before retaliating over the unauthorized access and illegal removal of two Democratic Party organizations' emails and documents.[65] Snowden concluded that "an escalation in the attribution game could get messy fast."[66]

CYBERATTACKS

Cyberattacks have long been part of the Russian shadow warfare strategy and arsenal. Moscow began cyberattacks on neighboring states like Estonia, Georgia, Kyrgyzstan, Lithuania, and Ukraine in 2007. By 2015, Russia was launching cyberattacks against further targets like Germany, Poland, the Netherlands, and the United States.[67] Initially, these were DDoS attacks that required considerable human participants. Then, the cyberattacks became more sophisticated, with more complicated malware. For instance, the governments of France, Germany, and the Netherlands, in advance of the presidential and general elections in 2017, agreed to share information as they braced for "influence operations," including the dissemination of illegally obtained emails and disinformation campaigns on social media. The goal of the influence and disinformation campaigns is to divide public opinion and ultimately undermine the concept of truth.[68] The targets that received the largest and most prolonged attacks were Ukraine, the United States, and Germany. What follows are some of the acknowledged cyberattacks by Russia.

An early cyberattack was launched in 2007 against Estonian government websites after a World War II war memorial was moved from a Tallinn park, called Liberators' Square during Soviet times, to a military cemetery. The Russian government was displeased by the removal of the Bronze Soldier commemorating Soviet troops that fought against Nazi Germany. The initial wave of cyberattacks were basic DDoS attacks that came from official structures in Russia, were hosted by Russian state computer servers, and instructions on how to carry out cyberattacks circulated in the Russian language on Russian websites. Politically motivated hacktivists in these DDoS cyberattacks utilized botnets—a network of internet-connected private computers infected with malicious software running one or more bots and controlled as a group without the owners' knowledge.[69]

Estonia was particularly vulnerable to DDoS cyberattacks because much of its government is run online. The Baltic state has a paperless e-government and holds parliamentary elections online.[70] Targets of the DDoS cyberattacks included the Foreign and Defense Ministries, newspapers and other media outlets, and financial institutions. In attempts to circumvent the havoc, some officials simply blocked access to the servers from outside Estonia in order to prevent the attacks.[71]

In June 2008, Russia conducted another DDoS cyberattack, this time in Lithuania. In this instance, the cyberattacks appear to be motivated by the Lithuanian government's decision to outlaw the display of Soviet symbols. In response, Russian hacktivists and privateers, presumably with the guidance of the Russian government, defaced Lithuanian government websites with Soviet symbols of the hammer and sickle, and red stars.[72]

Russian cyberattacks against Georgia, followed by military forces a few weeks later, were an early instance of full-fledged hybrid warfare.[73] Cyberattacks on the Georgian government sites began as early as July 20, 2008, with coordinated DDoS cyberattacks that overloaded and effectively shut down Georgian servers.[74] In the DDoS attack, malicious programs known as botnets were blasting streams of useless data at Georgian computers from hundreds of thousands of hijacked machines around the world, directing them to barrage Georgian sites. The bots effectively targeted the pages of Georgia's president, parliament, the Ministry of Foreign Affairs, news agencies, and banks. In addition to the DDoS cyberattacks, Georgian internet traffic began to be redirected through Russian telecommunications firms. A Russian-language site, Stopgeorgia, offered software to download for use by privateers in the DDoS cyberattacks. Using their signature tools and attack commands, a St. Petersburg-based criminal gang known as the Russian Business Network (RBN) launched cyberattacks from computers it is known to control. At one point, the parliament's website was replaced by images comparing Georgia's president to Adolf Hitler.[75]

Botnets were prepositioned in preparation for the attack and then activated shortly before Russian air strikes began.[76] While Russia baited Georgia with troop movements on the borders of the breakaway region South Ossetia, the bots were to attack. Then, in August 2008, after Georgia's pro-Western government sent troops into South Ossetia, the Russian military—land, sea, and air—invaded Georgia. In combination with this second cyberattack, the hybrid operation shut down Georgia's internal communications.[77] Russia prevailed in this hybrid warfare in just five days, with the Republic of South Ossetia declaring its independence, supported by Moscow. In the year after the war, Russian cyberattacks shut down social media platforms to commemorate the first anniversary of their success.[78]

The Central Asian republics were the last countries to leave the Soviet Union, and Moscow still has close ties with the republics, including Kazakhstan and Kyrgyzstan. However, Russian cyberattacks shut down two of the four Kyrgyz internet service providers with a DDoS cyberattack in January 2009. The cyberattacks appear to be part of a Russian effort to pressure Kyrgyzstan's government to remove a US military base set up after the September 11, 2001, attacks on the World Trade Center in New York from its national territory. After prolonged negotiations, the base was closed in

2014, and the Kyrgyzstan government later received $2 billion in Russian aid and loans. Three months following the Kyrgyzstan attacks, in April 2009, Kazakhstan's then-president Nursultan Nazarbayev released a published statement that was deemed critical of Russia. The media outlet that published the statement underwent a DDoS cyberattack that temporarily disabled its website.[79]

A Russian cyberattack against the Netherlands was conducted in October 2015. The goal of the cyberattack was to access Dutch government computers regarding a report concerning Flight MH17, which was shot down over eastern Ukraine in July 2014. It appears that the Russian government wanted to ascertain the data collected on the downing of MH17.[80] The Dutch Safety Board headed the investigation of the Malaysia Airlines' airplane destruction. The report concluded that the passenger plane was downed, and all passengers and crew were killed by a Russian-made missile fired from an area held by pro-Russian rebels.[81]

Ukraine has been one of the major targets for Russian cyberwarfare. For instance, a DDoS cyberattack, thirty-two times larger than the largest known attack in history, occurred while Russia seized control of Crimea from Ukraine in March 2014.[82] The intent of the attack was to take down telecommunications to prevent the Ukrainian forces in Crimea from communicating with each other and, perhaps more importantly, from communicating fully with the Ukrainian government in Kyiv while Russia introduced overwhelming physical forces. This was at least the second time that Russia combined cyber and kinetic warfare to create hybrid warfare.

The Russian government began using more sophisticated cyberattacks than DDoS cyberattacks in the aftermath of revelations about the use of sophisticated malware like Stuxnet in 2010 by the United States and Israel. During Ukraine's presidential elections in May 2014, first, there was a cyberattack that took down the country's Central Election Commission three days before the vote. The attack was designed to create chaos and hurt the nationalist candidate while helping the pro-Russian candidate, who lost despite these efforts.[83] Then, CyberBerkut, a pro-Russian privateering group, rigged the website of the Central Election Commission to erroneously announce the ultra-right presidential candidate as the winner. CyberBerkut is linked to the cyberattackers who later breached Democratic Party targets in America's 2016 presidential election.[84] Additional malware on the Central Election Commission's system appears to be traced back to the GRU's APT28/Fancy Bear unit.[85]

The barrage of cyberattacks accelerated in the autumn of 2015 and became even more sophisticated, using only the second discovered malware after Stuxnet[86] that targets industrial control systems. Russian government cyberattackers used malware to gain access to the power supply network in western

Ukraine, leaving 225,000 people without electricity.[87] Then, in December 2016, Russian malware shut down 20 percent of the electric power generated in the Ukrainian capital Kyiv. The malware, CrashOverride, targets industrial control systems with payloads that attack communication protocols and directly control switches and circuit breakers.[88] CrashOverride scans for critical components that operate circuit breakers and then opens the circuit breakers, which stops the flow of electricity. It continues to keep them open even if a grid operator tries to close them, creating a sustained power outage. Moreover, the malware erases the software on the computer system that controls the circuit breakers, forcing the grid operator to revert to manual operations, which means driving to the substation to restore power. With this malware, the attacker can target multiple locations creating outages in different areas simultaneously. Theoretically, the malware can be modified to attack other types of industrial control systems, such as water and gas. Electrum, which targeted the Ukraine electric grid in 2015, and Sandworm, which targeted US industrial control systems in 2014, may be the same group or two separate groups working within the same organization, but the forensic evidence indicates they are related.[89]

The Russian cyber strategy in Ukraine, according to Western analysts, is "to destabilize the situation in Ukraine, to make its government look incompetent and vulnerable"[90] as well as using Ukraine as a testing ground for perfecting new forms of cyberwarfare and cyberweapons. In addition to the blackouts and other cyberattacks, Russian disinformation flooded Ukraine's media, including persistent tales that portrayed it as a fascist state filled with anti-Semites, despite the fact that the country elected a Jewish president in 2019.[91] This particular brand of disinformation was meant to bring back glorious images of Russian victories in World War II and encourage national fervor. This Russian cyber strategy seems to be a model that was applied elsewhere.

The Russian cyberattacks on the 2016 US election seemed to follow a strategy similar to the one used in Ukraine, with a combination of cyberattacks that used a massive cyber disinformation operation combined with physical cyberattacks on election data and equipment. The official US review of the Russian cyberattacks initially focused on the disinformation operation, finding that "Russian President Vladimir Putin ordered an influence campaign in 2016 aimed at the US presidential election. Russian goals were to undermine public faith in the US democratic process, denigrate Secretary Clinton, and harm her electability and potential presidency. We further assess Putin and the Russian Government developed a clear preference for President-elect Trump. We have high confidence in these judgments."[92] The attacks were coordinated and run by the GRU and used the services of the privateer entity, CyberBerkut, which was used in so many other similar attacks.[93]

Much of the initial publicly available discussion focuses on the massive disinformation operation. The US Office of the Director of US Intelligence's report, "Assessing Russian Activities and Intentions in Recent US Elections," states:

> Putin and the Russian Government aspired to help President-elect Trump's election chances when possible by discrediting Secretary Clinton and publicly contrasting her unfavorably to him. All three agencies agree with this judgment. . . . Moscow's influence campaign followed a Russian messaging strategy that blends covert intelligence operations—such as cyber activity—with overt efforts by Russian Government agencies, state-funded media, third-party intermediaries, and paid social media users or trolls. . . . Russia's intelligence services conducted cyber operations against targets associated with the 2016 US presidential election . . . and that the GRU used the Guccifer 2.0 persona and DCLeaks.com to release US victim data.[94]

The US analysis argued for even more focus on how Russian disinformation—as generated by Russia's multiple state-owned platforms—was used to complement the full Russian influence campaign. Specifically, the US Senate report notes that open-source collection and reporting are traditionally used to support specific analytic assertions and should be used regarding RT and Sputnik's coverage of WikiLeaks releases of DNC information.[95]

The Russian cyberattacks on the 2016 US election had two physical components. First was the theft of emails and a campaign finance database through spear-phishing operations that reached as high as the campaign chair, John Podesta. The research into the Russian cyberattack group APT28/Fancy Bear corroborates US intelligence reports that Russia used unauthorized access into the DNC email system as part of a cyber campaign to interfere in the election. The CIA identified Russian officials who delivered the stolen DNC emails to WikiLeaks through third parties under orders by President Vladimir Putin. In some cases, the stolen documents traveled less directly from the GRU to WikiLeaks.[96]

The Russian cyberattacks on the 2016 US election had a second physical component, which has major ramifications. Russian cyberattacks directly went after voter databases and software systems. The US voter systems saw cyberattacks in 78 percent—or a total of thirty-nine—US states, many more than was originally reported. In Illinois, cyberattackers accessed software designed to be used by poll workers and actively tried to delete or alter voter data.[97] The Special Counsel Robert Mueller's report on Russian interference in the 2016 election reconfirmed that a spear-phishing "operation enabled the GRU to gain access to the network of at least one Florida county government."[98] Florida's senator Marco Rubio took it one step further, saying that Russian hackers not only accessed the state's voting system but were "in a position" to change voter roll data.[99] These cyberattacks on the voting system were confirmed in a report by the US Office of the Director of US Intelligence, warning that "Russian intelligence obtained and maintained access to elements of multiple US state or local electoral boards."[100] The cyberattacks against voter data and voting systems were further confirmed by NSA documents, outlining how the GRU targeted the computers of 122 local election officials prior to November 2016.[101] The unauthorized release of these NSA documents ultimately resulted in the guilty plea of Reality Leigh Winner, a young US Air Force veteran and intelligence contractor, for unlawfully retaining and disseminating defense intelligence information.[102] While she did not release documents to the GRU, she did release documents to the US press that confirmed that the NSA knew that the GRU had gained access to US election computers.

As part of this operation on the physical voting structure, the GRU executed a cyberattack against at least one US voting software supplier and sent spear-phishing emails to more than one hundred local election official days before the November election. The NSA acknowledges that the Russian government multipronged cyberattacks "specifically directed at U.S. and foreign elections,"[103] focused on parts of the system directly connected to the voter registration process, including a private sector manufacturer of devices that maintain and verify voter rolls. Some of the company's devices are advertised as having wireless internet and Bluetooth connectivity, which could have provided an ideal staging point for further malicious actions.[104]

Two other Russian cyber operations were part of this cyberattack on the physical voting structure. In one, Russian military cyber operators created an email account pretending to be a US election company from which they sent fake test emails offering "election-related products and services,"[105] such as those used in Florida. In the second operation, test emails were sent to addresses at the American Samoa Election Office, in prelude to launching another spear-phishing attack. The Russians appeared intent on "mimicking a legitimate absentee ballot-related service provider."[106] These particular attacks seem to be probing the weaker links in the US election process, which could discredit the reliability of the elections both domestically and internationally.

As mentioned earlier, energy companies in the United States were the intense subject of a series of Russian cyberattacks that have, in some cases, successfully broken into the core systems that control the companies' operations. Access was gained through spear-phishing operations. The cyberattacks began in late 2015 but increased in frequency in April 2017. Energy companies and other utilities are susceptible to cyberattacks that could be used for destructive purposes, such as limiting energy supplies that can damage the economy or disable key cities. Further, attacks on energy companies can coincide with physical attacks as part of a hybrid operation.[107]

In 2017, a Russian defense agency reviewed the source code of the cyber defense software used by the US military to guard its computer networks. The Hewlett Packard system, called ArcSight, is a cybersecurity software used by much of the US military, alerting analysts when it detects that computer systems have come under attack. ArcSight is also widely used in the private sector.[108] With full Russian access to the source code, this software no longer makes sense for use in the US military.

A Russian cyberattack using a cyberwarfare strategy similar to the one used against Ukraine and the United States was conducted against the Foreign Ministry of the Czech Republic in January 2017. Not only was it a cyberattack, but it also merged aggressive actions with cyberespionage, giving the attack its shadow warfare quality. Most senior diplomats' emails were accessed, prompting the Czech Republic's foreign minister to compare it to the cyberattack on the DNC. The attack retrieved electronic correspondence concerning the Czech Republic's relations with its NATO and European Union allies.[109] This sophisticated cyberattack came amid an active Russian disinformation campaign in the Czech Republic, first around a US effort to locate advanced missile systems on its territory and then to focus on the social issues that could impact the presidential election.[110]

There is less information available on a Russian cyberattack in 2017 on the Polish Foreign Ministry, although it appears to resemble the attack on the Czech government. The malware used in the cyberattack was sophisticated enough to be disguised as legitimate software. The Polish Foreign Ministry asserts no classified information was compromised as only the internal system was affected by the attack.[111] This use of more sophisticated cyberweapons in attacking Poland came after thousands of DDoS attacks from 2011 to 2014 against the government and financial targets.[112] Attribution of these earlier DDoS cyberattacks in Poland was claimed by CyberBerkut. CyberBerkut is the same group associated with Russia that attacked the Ukrainian elections in 2014, the German government in 2015, and the US elections in 2016.

The cyberattacks on Germany follow the same strategy that was used in Ukraine and the United States—initiate the cyber operation with a massive disinformation campaign and then pair it with cyberattacks on physical structures like governmental emails, political parties, and infrastructure. Both Russian privateers like CyberBerkut, and Russian GRU units like APT28/Fancy Bear and APT29/Cozy Bear, are used.

In January 2015, CyberBerkut commenced a two-day DDoS cyberattack on German government computers to coincide with a visit of the Ukrainian prime minister. Relations between Ukraine and Russia experienced unprecedented alienation since 2014, with the Euromaidan revolution, the annexation of Crimea, and the outbreak of war in Donbas. While Russia argues it is providing economic and political support to the Russian-speaking populations of Crimea and eastern Ukraine from a supposedly failed Ukrainian state, Germany and the West have backed treaties and sanctions to move Russia out of Ukraine to help unify the country[113] and have considered Ukraine's membership in NATO—right on Russia's border.[114] In April and May 2015, the GRU's APT28/Fancy Bear conducted a sustained cyberattack over several weeks on the Bundestag, the national parliament of the Federal Republic of Germany. German intelligence agencies concluded that these cyberattacks were directly authorized by Russian president Vladimir Putin. The distinctive Russian character of this attack is exposed by the fact that not only was it a cyberattack, but it also merged aggressive actions with cyberespionage. It was the BfV, Germany's domestic intelligence service, that asserted the cyberespionage sought information on the workings of the Bundestag, NATO, and German leaders, including high-ranking politicians in Chancellor Angela Merkel's Christian Democratic Party.[115] The cyberattack infected a network of more than 5,600 computers and 12,000 registered users—including Merkel's office—with malware and stole 16 gigabytes of data. The attack was so severe that the entire Bundestag network was taken offline for four days.[116]

Then, in December 2016, the BfV warned of growing evidence of Russian cyberattacks and disinformation campaigns to influence the German parliamentary elections in September 2017 and to destabilize German society. Angela Merkel, who supported sanctions against President Putin's personal associates after Russia annexed Crimea, was specifically targeted.[117] Two foundations affiliated with Germany's ruling coalition parties were also attacked.

The disinformation operation of the Russian cyber campaign against Germany in 2017 had several components. The three major outlets for the disinformation campaign are RT Deutsch, Sputnik Deutsch, and NewsFront Deutsch. The distribution of disinformation was aided by bots and by human networks, often connected to pro-Russian factions, as well as far-right (conspiracy) media outlets or anti-migrant groups.[118] The best-known case in the disinformation campaign was the media flare-up surrounding a phony story about a Russian-German girl who had reportedly been raped by Arab migrants, intended to manipulate German public opinion. Germany's leading role in the Ukraine crisis, along with Angela Merkel's consequent position on sanctions against Russia, made the German government a core target of Russian disinformation.[119] The German media has responded by adding fact-checking and investigative capabilities, but with limited effect.[120]

The German government's response has validated a tough stance toward Russia, substantiated the need for increasing Germany's defense budget, enhanced its commitment to NATO, and garnered respect for the rarely popular German intelligence services. Part of the response involved allowing authorities to wipe servers used by APT28/Fancy Bear and APT29/Cozy Bear to conduct the cyberattacks. Chancellor Merkel, as well as her former and current foreign ministers, announced that the strategic relationship with Russia is over.[121] The Russian-German political relationship remains strained.[122]

European governments—especially France, Germany, and the Netherlands—warned of cyberattacks in the wake of their 2017 presidential and general elections. With the familiar strategy, the cyberattacks included disinformation campaigns, cyberespionage that obtained emails and other data, and attacks on political parties and government institutions. One example of this was France. Illegally obtained emails—emails that discussed all aspects of the presidential campaign—were distributed for maximum harm. Disinformation was spread on social media using bots and through Russian state-owned foreign-language news services, Sputnik and RT, in attempts to disrepute political leaders and political parties.[123] Yet Russian interference succeeded neither in interfering with the 2017 French presidential election nor in antagonizing French society. In the spring of 2017, an orchestrated disinformation campaign against Emmanuel Macron's presidential campaign began. The so-called Macron Leaks—a combination of real emails and forgeries—were released online just two days before the final round in the vote. France was at an advantage because it was targeted after cyberattacks and disinformation campaigns were launched in the Netherlands, the United Kingdom, and the United States. In France, administrative, independent, and nonpolitical authorities provided technical and politically neutral expertise to ensure the integrity of the electoral process.[124]

In 2017, APT29/Cozy Bear carried out cyberattacks on the Norwegian Foreign Ministry, military, intelligence, and other institutions through nine email accounts that were targeted by spear phishing. In addition, the cyberattack targeted the Norwegian Radiation Protection Authority, a school, and the parliamentary group of the Labor Party. Norway, a NATO member and Russia's neighbor, normally enjoys good relations with Moscow, but the relationship has grown strained recently. The cyberattacks may be related to the September 2017 Norwegian elections or to Norway's participation in the EU's economic sanctions against Russia over the Ukraine crisis, or to the deployment of approximately 300 US soldiers on Norwegian soil.[125]

The increase in cyberattacks originating in Russia from 2007 onward is largely designed to be influence operations or to be used in conjunction with powerful military force. In a classic influence operation, nothing needs to be physically manipulated. It is gathering information and spreading disinformation to foment social discord or create payback for political decisions or influence national elections.[126] Sometimes, however, Russian cyberattacks coincide with military action, initially in Georgia but more recently in Ukraine with the annexation of Crimea and intervention in the east. Influence campaigns are less expensive than military action. Cyberattacks coordinated with kinetic military force are more effective.

CONTROLLING CYBERSPACE

The Russian government believes that it can control the most necessary aspects of cyberspace. The concept of controlling cyberspace is addressed in the December 2014 Military Doctrine of the Russian Federation.[127] Under Part II, Section 12, "Military Risks and Military Threats Encountered by The Russian Federation," the main risks are listed as the "use of information and communication technologies for the military-political purposes to take actions which run counter to international law, being aimed against sovereignty, political independence, territorial integrity of states and posing threat to the international peace, security, global and regional stability."[128] Then, section 21, "Main Tasks of the Russian Federation Regarding Deterring and Preventing Military Conflicts," outlines that controlling cyberspace is needed "to create conditions to reduce the risk of using information and communications technologies for the military-political purposes to undertake actions running counter to international law, directed against sovereignty, political independence or territorial integrity of states or threatening international peace and security, and global and regional stability."[129]

A central aspect of controlling cyberspace is domestic control. Russia is testing whether it can exert almost total control of domestic cyberspace by disconnecting from the global internet. Cutting itself off from the World Wide Web at this point is a difficult task that could have unintended consequences. If anything, the Russian endeavor to be able to disconnect illustrates just how entangled—and strong—the global internet has become.[130] As mentioned earlier, the end goal is for Russian authorities to implement a cyberspace traffic filtering system like China's Great Firewall, but also to have a fully working domestic cyberspace in case Russia needs to disconnect, presumably as the result of or in anticipation of cyberattacks.[131] However, while China built its domestic internet system while the global internet was just emerging, Russia is faced with constraining an existing system that has grown without many restraints over the past three decades. This is clearly more difficult. For instance, Roskomnadzor made an ambitious, if disastrous, effort to ban Telegram, the messaging service. Telegram was accused of failing to comply with FSB requests to share user data. In response to the attempt to shut it down, Telegram's Russian founder rerouted its traffic through cloud hosting services, forcing Roskomnadzor into a game of whack-a-mole that saw the regulator temporarily take down more than sixteen million IP addresses, including its own website, while having little effect on Telegram.[132]

Within the central aspect of domestic control, the economic consequences of restricting internet traffic are not a high priority. The current economic climate, the fall in oil prices, and international sanctions have had a negative effect on the Russian economy overall. These concerns have resulted in a weakening currency, stoking inflation, squeezing household incomes, and limiting the potential for e-commerce.[133] Russian e-commerce market does exist: a 2018 Morgan Stanley study predicted its nearly threefold growth between 2018 and 2023.[134] Nonetheless, e-commerce still only represents 3 percent to 4 percent of the total Russian retail market.[135] The limited size of the market is clear from the small sales volumes of Russian online retailers in comparison with major foreign economies.[136] Part of the issues regarding e-commerce in Russia revolves around governmental decisions to collect information regarding online purchasers and restricting the methods by which to pay for online goods. For instance, bank cards or credit cards are not as ubiquitous in Russia, and when they are used, card details must be stored by the card provider and be available to the government. Moreover, the Russian Post—the biggest national postage service provider—suffers from severe delays in product delivery and contributes to customer dissatisfaction. Finally, the potential for cyberattacks damaging the Russian economy is more concerning when it comes to vulnerable industrial controls in the economically crucial oil and gas sectors rather than e-commerce.

Controlling cyberspace similarly focuses on military operations in and through cyberspace. The Kremlin is certain that Russia is locked in an ongoing, existential struggle with forces that are seeking to challenge its security in cyberspace, bolstered by general Western opposition to is cyber and noncyber policies in Ukraine and by specific cyberattacks on its infrastructure like its electric grids. Cyberspace—specifically the ability to move cyberweapons through it, conduct cyberespionage, and distribute information and disinformation—is viewed as both a threat and an opportunity. In keeping with an established Russian view of battling persistent threats during the era of shadow warfare, the Russian government views the struggle within cyberspace to be a constant and unending tug of war of defense and offense, with the West in particular. This suggests that the Russian government has a relatively low bar for employing cyber operations in ways that other states are likely to view as offensive and escalatory in nature.[137]

Perhaps the most pertinent perspective regarding controlling cyberspace is the Russian government's international stance. As mentioned in the previous chapter, there are two main camps when it comes to controlling cyberspace. One camp of countries holds that the issue of cybersecurity can be resolved only by a treaty process, and Russia is an early leader in this camp. The other camp argues that any undesirable state uses of cyberspace can be dealt with adequately under existing international law. The fault lines between the two camps resemble those of the Cold War, with Russia and China belonging to the camp of sovereign internets protected by treaties and the United States and much of the West belonging to the camp of a relatively unfettered international internet.[138]

The call for norms in cyberspace has its roots in a Russian-led arms control resolution dating back to 1998. This early resolution focused on mitigating threats from cyberweapons and information wars while pushing for the ability to retain control over domestic cyberspace. This laid the groundwork for the Russian view of cyberspace as needing to be controlled.[139] Russia created momentum and gathered international support to assist its camp's calls for the creation of new cyberspace regulations, for a cyber treaty, and for governmental control of domestic cyberspace, especially the flow of information. Russia has striven to bring the issue of the control of cyberspace into the First Committee of the United Nations, which is tasked with nuclear disarmament, nonproliferation, arms races, and illicit arms trade.[140] As part of this push to have cyberspace control under international treaty and regulation, Russia built a counternarrative to the Chinese view of cyberspace control as information traffic control and to the American defense of the freedom of information, and instead focuses on controlling cyberspace by means of a new and specific international legal regime with strong sovereign controls. The international camp in which Russia is a leader might well have attempted to achieve just that, given the emphasis of Russia's submissions to the First Committee between 1999 and 2003. Several countries shared the Russian view on the advisability of an international arms control regime with regard to cyberweapons and cyberwarfare.[141]

Between 2005 and 2009, Russia shifted its focus, not from concretizing cyber rules and norms into a treaty but toward doing so at a regional level rather than exclusively on the international stage. This resulted in several regional treaties on the control of cyberspace. First, the Agreement on Cooperation in Ensuring International Information Security between the Member States of the Shanghai Cooperation Organization was concluded between People's Republic of China, Russia, Kazakhstan, Kyrgyzstan, Tajikistan, and Uzbekistan on July 16, 2009. The agreement, aimed at ensuring "international information security," deals specifically with cooperation on confronting the following threats to international cybersecurity: (1) development and use of information weapons, preparation and conduct of information warfare; (2) information terrorism; (3) cybercrime; (4) use of a dominant position in the information space to the detriment of the interests and security of other states; (5) dissemination of information prejudicial to the sociopolitical and socioeconomic systems, spiritual, moral, and cultural environment of other states; and (6) threats to the secure and stable functioning of global and national information infrastructures that are natural and/or manmade.[142]

Then, on September 22, 2011, Russia tabled an international cybersecurity treaty, the Convention on International Information Security (sometimes short-handed as the Concept Convention), which at the time was distributed primarily by Russian embassies and diplomatic representations, via the Ministry of Foreign Affairs of the Russian Federation website, to make it freely available to any interested party.[143] Next, in 2013, a treaty on cooperation was concluded among the members of the Commonwealth of Independent States—Russia's near neighbors—to improve cybersecurity.[144] Finally, in 2011 and 2015, Russia again utilized the regional members of the Shanghai Cooperation Organization to submit to the United Nations an international draft treaty aimed at facilitating international consensus on norms and rules guiding the behavior of states in cyberspace.[145] It is unlikely that Russia will change its push for this approach to cybersecurity since this camp reflects Moscow's best interests.

A 2017 draft of a Russia's fifty-four-page proposal for a "United Nations Convention on Cooperation in Combating Information Crimes" includes seventy-two proposed articles, covering the collection of cyberspace traffic by authorities, codes of conduct for cyberspace, and joint investigation of malicious activity. The Russian government's proposed treaty would enhance the ability of Russia and other members of the camp to control their domestic cyberspace and to gain access to communications in other countries. This draft is another part of Russia's push over the past decade to shape the international regime of cyberspace.[146]

Controlling cyberspace, both domestically and internationally, has central importance to the Russian government. A Russian website, Pravda,[147] published an interview in 2013 with the chairman of the State Duma Committee on Information Policy, Information Technology and Communications, Alexei Mitrofanov, in which he outlined the Russian belief that technologically controlling cyberspace was not only possible but also necessary. According to Mitrofanov, "Everything can be controlled. . . . Technically there are no problems. . . . IT was originally started at the US. . . . [Russians] must admit that IT includes military products, that is, it is declassified military, and the internet is military. And it is very sad that the Russian military in 1991 chose to lease premises. . . . *We had the best computers*. Yes, we did. But we did not turn it into business."[148] The Russian government is determinedly creating a cyber environment in which Russia is an important actor and is eliminating any reservations about controlling cyberspace. Control of cyberspace is needed to project power and defend the domestic economy, the military, and the government.

SHADOW WARFARE POLICY

Fundamentally, Russia's shadow warfare policy is guided by political, economic, and social realities. Russia agrees with the United States that cyberspace is an incipient battleground. There the agreement ends. While the United States is concerned primarily with threats to technology and economic well-being, Russia is concerned about cyberactivity that threatens interference in Russian sovereign political affairs.[149] Therefore, Russia favors an international treaty along the lines of those negotiated for chemical weapons. From a Russian perspective, the absence of a treaty is permitting potentially dangerous consequences, including social and political stability. The Russian proposed treaty called for three measures: a ban on a state secretly embedding malware that could be later activated from afar in the event of war, bar the application of humanitarian law (that the West wants to use for global internet norms); prohibition of deception in cyber operations to deal with the challenge of anonymous cyberattacks; and broader international government oversight of the internet.[150]

Focusing on domestic political stability, the cyberwar policy approved by President Vladimir Putin in 2013 outlines cyberattacks as a major threat to international security and suggests countering it with a special international body and international behavior code to manage cyberspace, which was prepared, in part, as a reply to the International Strategy for Cyberspace approved by the United States in 2011. The main threats mentioned in the policy were both focused on domestic political stability: an "informational weapon used for military-political, terrorist and criminal ends" and attempts of "intervention into other nations' internal affairs."[151] The proposed international body and code of norms to manage cyberspace focused on limiting the potential to interfere with other states' domestic politics after events demonstrated the potential of online social networks to impact domestic politics, such as launching street protests—an area where Russia has long suspected instigation by the United States and the wider West through social media and the internet.

The Russian government asserts that its policy promotes more peaceful politics than the US approach, which equated cyberattacks to kinetic warfare and declared that the US military would react to cyberattacks using all means necessary—including nuclear weapons. The Russian policy emphasizes the strengthening of international cooperation and preventive regulative measures that would stop potential cyberattacks. The proposed measures include the approval of the UN convention on international cybersecurity and developing "internationally accepted rules of behavior in cyber space." Russia also wants to develop an international system to manage the internet and to impose an international law that would "prevent the proliferation of the informational weapons." Russia wants to make possible the political control of the internet by all states.[152]

Second, Russia's top strategic body, the Security Council, and the main security agencies are now expected to provide President Putin with suggestions on particular measures to enforce shadow warfare policy. For instance, the 2013 bilateral agreement between Russian and the United States on the prevention of cyber incidents developing into interstate conflict is described as a typical example of positive cooperation.[153] This more institutionalized approach was designed to facilitate the regular exchange of practical technical information on cybersecurity risks to critical systems by sharing threat indicators between Russian and American computer emergency readiness teams. On a continuing basis, these two institutions will exchange technical information about malware or other malicious indicators—appearing to originate from each other's territory—to aid in proactive mitigation of threats.[154]

Finally, when it comes to national and ideological principles that shape policy, in an ongoing process to establish international norms, Russia's cyberwar policy that underlies its shadow warfare policy deviates from that of the United States and, ultimately, of China. The main points of departure focus around the ideological underpinning of the Putin government's war doctrine and the "sometimes necessary" view of war by the Russian Orthodox Church. The Putin administration's approach to cyber policy relies on earlier Soviet policy and practice. This is most evident in the Russian reliance on weaponizing information and using *kompromat*—compromising material—gleaned from cyberespionage and other sources. President Putin, starting in his career with the Soviet KGB, used *kompromat* against Russian politicians for decades. The FSB deploys this tactic against foreigners, both in Russia and abroad, gathering *kompromat* in case it will come in useful one day. For example, a US diplomat is purportedly shown in a video with a prostitute before he was a diplomat in the US Embassy in Moscow.[155] One purpose of the cyber-released and cyber-obtained *kompromat* is to undermine governments and government officials through weaponized information and to regain Soviet-level status on the world stage for Russia.

The Russian Orthodox Church, often in direct cooperation with the Putin government, also frames Russian cyberwar policy. This reliance on the Orthodox cover to create war policy is supported by a popular Russian nationalist thinker, Aleksander Dugin. Dugin's *The Fourth Political Theory*[156] has implications for cyberwar policy, with his assertion that Russia, in opposition to Western modernity principles, stands on the side of tradition. Russia and the Russian Orthodox Church, according to Dugin, must go their own way, refuse universalism, and insist that the different peoples of the world must rediscover and rejoice in their own diverse cultures and traditions to create a multipolar world. Dugin, who purports to be an Orthodox Christian, is not at all concerned with whether these traditions are authentic. The Russian Orthodox Church, insisting that it is not bound by Western notions of separation of church and state, is recreating a policy of modern warfare and cyberwarfare. While recognizing war as evil, the church does not prohibit hostilities if the security of its neighbors and the restoration of trampled justice are at stake. When that is the case, war is necessary, according to Section VIII ("War and Peace") of the Department for External Church Relation's document, "The Basis of the Social Concept." The Russian government has relied on the church as a unifying force in a religiously diverse country, and, in turn, the role and influence of the church in the government and the military have been steadily growing, to the point that Russia's Defense Ministry constructed its own cathedral.[157]

Cyberwarfare policy in Russia is both a reaction against the cyberwarfare conducted by other actors and an active policy to undermine other governments in order to create a multipolar world. Russia's push for international control of cyberspace is an attempt to wrestle primary control from the United States and the West in order to diversify the "ownership" of the international cyberspace. This allows Russia to be a more important actor in cyberspace, denies the West primary influence in the realm, and pushes the world order more definitively toward the multipolarity that Russia believes is in its national interest.

Chapter 4

Cyber China

This chapter looks at the national policies and the underlying doctrines of the People's Republic of China regarding power projection, national security, and strategic planning. Also, of significance are examples of cyberattacks, such as who is being attacked, how, and why. While China follows a more extensive industrial cyberespionage policy than the other two major powers in shadow warfare, the move toward a more military focus on cyberspace operations is similar to the United States and Russia. One comparatively unique aspect of China's shadow warfare concerns is its global leadership approach to national cyber governance. Specifically, China promotes sovereignty-based governance schemes that allow states to nationally regulate cyberspace. In part, China prefers national sovereignty-based cyberspace governance due to disproportionate Western dominance in shaping the future of global cyberspace. In part, China's administration is aware of the negative potential threat of uncontrolled information and the positive use of cyberspace as a strategic weapon to achieve an asymmetric advantage.[1] With the largest online population in the world—854 million internet users as of June 2019, mostly on their cell phones[2]—China is openly declaring its place as a cyber power along with the United States and Russia, as well as other players, determinedly assuming a leadership role in this new form of warfare and power projection.

China discusses its own emphasis on cyberwar capabilities in several official documents, including the special emphasis in the 2016 National Cyberspace Security Strategy on cyberspace as a new territory of national sovereignty: "Cyberspace has become a new field of human activity that is as important as land, sea, sky and space. The expansion of national sovereignty extends to cyberspace, and cyberspace sovereignty has become an important part of national sovereignty. Respecting cyberspace sovereignty, maintaining cybersecurity, seeking common governance, and achieving win-win results are becoming the consensus of the international community."[3]

While the introduction of the 2016 National Cyberspace Security Strategy mentions many positive aspects of cyberspace for China and for the international community, it also discusses serious challenges as well mentioning political, economic, social, and cultural security. Moreover, the massive worldwide WannaCry ransomware attack of 2017 had a vast impact on China's fears of foreign-based cyberattacks.[4] As discussed earlier on, the malicious backdoor software that cyberattackers relied on to develop the WannaCry ransomware was created by the NSA and later stolen by a secretive group known as the Shadow Brokers,[5] with Edward Snowden writing that the "circumstantial evidence and conventional wisdom" suggested Russia was behind the theft.[6]

In the 2019 *China's National Defense in the New Era* white paper, Beijing reiterated its focus on the ability of cyber technology to impact the national development and the national security in the era of shadow warfare:

> Cyberspace is a key area for national security, economic growth and social development. Cyber security remains a global challenge and poses a severe threat to China. China's armed forces accelerate the building of their cyberspace capabilities, develop cyber security and defense means, and build cyber defense capabilities consistent with China's international standing and its status as a major cyber country. They reinforce national cyber border defense, and promptly detect and counter network intrusions. They safeguard information and cyber security, and resolutely maintain national cyber sovereignty, information security and social stability.[7]

China's 2016 strategy takes the intellectual high ground and is less fearful than the 2019 report's emphasis on threats. The National Cyberspace Security Strategy states that "cyberspace opportunities and challenges coexist, and opportunities outweigh challenges."[8] This is followed by the goal that,

> Guided by the overall national security concept, we will implement the development concept of innovation, coordination, green, openness, and sharing, enhance risk awareness and crisis awareness, coordinate the two major domestic and international situations, and coordinate the development of two major events, actively defending and responding effectively. Promote cyberspace peace, security, openness, cooperation, orderly, safeguard national sovereignty, security, development interests, and achieve the strategic goal of building a network power.[9]

The emphasis on cyberspace sovereignty echoes through the primary documents, illustrating the strong preference for state sovereignty.[10] The increasing emphasis on cyber threats echoes similar statements and activities from the United States and Russia.

INSTITUTIONS AND INDIVIDUALS

The principal institutions and individuals taking a central role in creating China's evolving shadow war policies have evolved over recent years as its cyber policy matures. One of the elements of this evolution includes a building and reorganization of the principle institutions into a more sophisticated network. This cyber network includes at least three components: first, a specialized military network of cyberwarfare forces; next, teams of network cyberwarfare specialists in government civilian organizations, including intelligence and security; and, finally, nongovernmental institutions that engage in cyberattack and cyber defense, including its civilian computer software and hardware industries.[11]

As an example of this evolving structure, the military network's advanced persistent threat group, APT1/ Unit 61398—the Chinese military group notorious for cyberespionage—appears to be largely out of business, with its cyber operators dispersed to other military, civilian, and intelligence units. The change is part of President Xi Jinping's broad effort to bring institutions that create cyber policies and strategies under his control. Institutions now directly under his control include the Central Internet Security and Information Leading Group, the technical strengths of the Ministry of Industry and Information Technology (MIIT), the law enforcement efforts of the Ministry of Public Security, the espionage skills of the Ministry of State Security, and the cyberwar sponsors within the Chinese military.[12]

With the 2014 creation of the Central Internet Security and Information Leading Group, the institution in charge of designing the national cyber policy, President Xi began the transfer of responsibility to define China's cyber operations to him personally.[13] This creation of a presidential-led institution to guide cyber policy resulted in the release of a new Five-Year Plan on cyber operations by the State Council of the Republic of China in December 2016, which states that "a five-year plan on China's national informatization (2016-2020) . . . will put more resources into the development of cutting-edge information technology, including 5G wireless systems, IPv6, smart manufacturing, cloud computing and internet of things. The plan sets a goal of authorizing 15.3 trillion patents in the information industry."[14]

Other goals set forth in this Five-Year Plan of cyber capabilities include the May 2020 completion of a satellite-based global positioning system, additional public integrated national databases from the government, academic institutions, and the public sectors, and expanded investment in information infrastructure for rural and remote regions. Another goal emphasizes the importance of a "smart government" in China, with a unified online system, integrating information and services from different departments and regions, and with the ability to deal with 80 percent of the paperwork online.[15] In addition, by augmenting the speed of the domestic internet and lowering costs, the government plans to connect the internet industry with manufacturing and agriculture, expanding e-commerce trade volume in 2020. A final goal of the plan is focused on domestic cybersecurity, setting up risk alerts and an emergency mechanism, and limiting telecom fraud.[16]

These national-level plans are carried out by a variety of governmental institutions. One of the major institutions is the MIIT, established in March 2008 in an effort to centralize cyber technology development.[17] This centralization through MIIT includes the redesigned Commission for Science, Technology and Industry for National Defense (COSTIND) and the Ministry of Information Industry. MIIT also includes the State Administration for Science, Technology and Industry for National Defense (SASTIND), which regulates the internet as well as other information entities. MIIT sets standards, holds exercises, inspects network security, and coordinates information and telecom security within the government. The primary duty to respond to cyberattacks rests with the nongovernmental technical center National Computer Network Emergency Response Technical Team/Coordination Centre of China (CNCERT), established in 2002. For its part, MIIT governmentally supports CNCERT's work by building malware and vulnerability databases, finding malicious information protocols and domain name providers, and guiding CNCERT's engagement in international cooperation.[18]

Other Chinese institutions also conduct important cyber policy. The Ministry of Public Security, the principal national law enforcement and security institution, investigates cybercrime and is responsible for critical infrastructure protection and development work through a wide network of research labs. The Ministry of Public Security is also responsible for overseeing the commercial products used by the government and controlling all commercial information security companies. Importantly, the Ministry of Public Security operates the Great Firewall of China—the software that keeps China under national sovereignty-based cyberspace governance and enables extensive censorship and domestic cyberespionage.[19] The Great Firewall came into existence within the first decade of the internet coming to China, so by the late 1990s it was firmly in place. China is not the only country that censors and attempts to control its domestic internet—with countries like Sudan, Kazakhstan, Brazil, Bangladesh, and Zimbabwe leading the global decline in internet freedom—but it is one of the most capable of achieving it. Agencies that track internet censorship, like Freedom House, confirmed China's status as the world's worst abuser of internet freedom for 2015–2019.[20] Although the techniques have varied, digitized, and intensified over time—the Great Firewall was loosened during the 2008 Beijing Olympics—it relies on banning key words, private enforcement by domestic institutions under threat of repercussions for failure, and by hiring a large workforce to manually scan the internet to catch rule breakers. While not impervious to penetration, particularly in an era in which malicious software can jump into cyber networks without a physical connection, the domestic cyberspace in China does uniquely use computer code written in Chinese, as opposed to the more ubiquitous English language code, which makes it less vulnerable to random cyber threats. Under China's Cybersecurity Law—a series of laws and regulations meant to guide China's domestic internet use by businesses and providers—the Ministry of Public Security is the main institution tasked with protecting domestic cybersecurity and combating cybercrime.[21]

The Chinese Ministry of State Security (MSS)—the primary national civilian intelligence agency, similar to the CIA in the United States and Russia's FSB—conducts counterespionage, counterintelligence, foreign intelligence, and domestic intelligence. Its efforts originally focused on countering terrorism—or the often-mentioned trilogy of separatism, terrorism, and religious extremism. The MSS' estimated cyber capabilities have grown significantly in order to collect political and economic data on foreign governments, nongovernmental organizations, and private citizens.[22] For instance, MSS cyber operatives broke into networks of eight of the world's biggest technology service providers in an effort to steal commercial secrets from their clients in a yearlong operation.[23] Under the 2017 National Intelligence Law, the MSS—along with other intelligence authorities—has broad powers to conduct various types of cyberespionage activities both inside and outside of China and to monitor and investigate foreign and domestic individuals and institutions.[24] A small number of the groups that employ more complex techniques and are effective at maintaining persistence in targeted networks have been attributed to the MSS. It appears that these groups include advanced persistent threat groups APT3 and APT10, both of which have been attributed to cyber operatives working on behalf of the MSS.[25]

As the military is often the main institution to conduct cyberwarfare, China's commitment to cyberwarfare has also impacted modernization and reform of its armed forces. Within the People's Liberation Army (PLA)—the official name for the entire Chinese military—the most notable institution within the organizational restructuring is the newly created military Strategic Support Force. The Strategic Support Force consolidated previous military branches that oversaw information, space, intelligence, surveillance, and reconnaissance support and created a separate military branch equal to that of the air force, navy, and army. The Strategic Support Force is a new type of combat corps for safeguarding national security and an important driver for the growth of new combat capabilities. It comprises supporting forces for battlefield environment, information, communications, information security, and new technology testing. In line with the strategic requirements of integrating existing systems and aligning civil and military endeavors, the Strategic Support Force is seeking to achieve major development strides in key areas and accelerate the integrated development of new combat forces so as to build a strong and modern strategic support force.[26]

What the establishment of the Strategic Support Force during the armed forces reforms of 2015 makes clear is that the military views cyber dominance and cyberattacks as not just a supplementary or supportive part of its war-fighting capabilities but an integral necessity and a crucial component of future Chinese defense and power projection.[27] Now, this is also clear in practice, with *China's National Defense in the New Era* asserting that "the PLA Strategic Support Force (PLASSF) has made active efforts to integrate into the joint operations systems [and hybrid warfare]. It has carried out confrontational training in new domains and trained for emergencies and combats."[28]

The military also maintains ties with other institutions such as research universities[29] and the public sector.[30] The Chinese military maintains a network of universities and research institutes that support cyberwarfare-related education either in advanced degree programs or in specialized courses.[31] The restructured military universities supporting this approach include the reorganized National Defense University and the National University of Defense Technology. China's armed forces have established the Central Military Commission Steering Committee on Military Scientific Research and reorganized the Academy of Military Sciences (AMS) and the services' research institutes. Thus, the military's scientific research forces have been rebalanced with the AMS as the lead, the research institutes, and the various arms of the services as the main focus, and the research components in educational institutions and the troops as supplements.[32]

China, like many countries, initially turned to its civilian internet technology workforce, but this strategy, too, was modified as Chinese cyberwarfare strategy matured. In the early days, between 1999 and 2004, China's privateers gained notoriety for their willingness to engage in large-scale politically motivated distributed denial of service attacks (DDoS), data destruction, and web defacements of foreign networks. While initially encouraged, this sentiment changed largely due to concerns about domestic consequences of critical computer users, with editorials in the official media[33] suggesting that cyberattacks and disinformation activity would not be tolerated.[34] Nonetheless, traditional privateers may still offer unique skill sets and may have a niche role for military or state intelligence collection.[35] Some evidence suggests that a relationship exists between Chinese privateers who specialize in maliciously violating computer security and Chinese government operators responsible for network intrusions. This leads to limited recruiting from among Chinese privateers,[36] similar to what occurs in the United States and Russia.

CYBER STRATEGY

China's cyber strategy relies on three major components. First are the purely military components that include cyber connectivity of their own military forces, the ability to disrupt the command and control systems of an adversary's military, the ability to fend off international malware and withstand DDoS attacks, and a capability to impose similar cyberattacks on others. This is a dominant and long-standing aspect of China's cyber strategy. Second are the economic aspects of the strategy that cover both domestic economic development and international cyberespionage on economic targets. This, too, has been a long-standing aspect of China's cyber strategy. Finally, China's cyber strategy focuses on domestic cybersecurity, which includes a national internet, domestic surveillance, and the protection and censorship by the Great Firewall. These military, economic, and domestic components have been present throughout China's cyber strategy over the decades.

The Chinese military cyber strategy clearly shifted in 2014–2015 to the version recognizable now. As part of this shift, *The Science of Military Strategy*, released in the spring of 2015, acknowledges for the first time—despite outsiders' assertions for years—that China is a cyber power with a network of cyberattack forces. The 2015 document also declares that the defensive and offensive cyber forces are divided into a specialized military network warfare forces, teams of network warfare specialists in government civilian organizations, and entities outside of the government that engage in network attack and defense, including the civilian information technology industry.[37] Similarly, another official 2015 document, *China's Military Strategy*, asserts that "China will devote more efforts to science and technology in national defense mobilization, be more readily prepared for the requisition of information resources, and build specialized support forces. China aims to build a national defense mobilization system that can meet the requirements of winning informationized wars and responding to both emergencies and wars."[38] This new openness about the need for strong cyber forces and the integration of civilian specialties into national defense was a definitive shift that portended the aggressive strategy shifts of 2019.

This 2014–2015 cyber military strategy shift was part of an overall military modernization and reform that occurred at this time. This reform was announced at the Central Military Commission Reform Work Conference in Beijing in November 2015. At the conference, the Central Military Commission—China's highest military body—outlined a strategy to implement these reforms by 2020.[39] In addition to the reforms that consolidated the previous seven military regions into five theater commands and set up new joint units, the strategy is to strengthen the forces in charge of cyberwarfare as well as missiles and space forces, "which are growing in importance in today's military conflicts."[40]

However, it is a mistake to see the Chinese cyber strategy in a merely military context. Bill Priestap, the former assistant director of the counterintelligence division of the FBI, in a statement before the Senate Judiciary Committee in 2019, asserted: "The Chinese government understands a core lesson of the Cold War between the United States and the Soviet Union: economic strength is the foundation of national power. The competition between the United States and China will be greatly influenced, if not ultimately decided, on the strength of our economies."[41] China's cyber strategy contains economic components that focus on generically enhancing Chinese economic development, influencing China's role in the world economy, and building a globally competitive military.

The Chinese economic cyber strategy also shifted in 2014–2015. A FireEye study concluded that as early as 2014, around the time of the indictment of Chinese military officers and cyberattackers in the United States for economic cyber theft, the Chinese government was modifying its approach to cyber operations.[42] Economic development aided by cyberespionage is still part of China's cyber strategy, but it is less likely to be conducted by the military cyber forces and more by civilian cyber operations. As China has grown into a major global economy, economic-based cyberespionage has become more targeted as well. For instance, the United States accused Chinese military officers of accessing US nuclear, metal, and solar companies to steal trade secrets, including Alcoa Inc, United States Steel Corporation, Westinghouse Electric Company, and the US subsidiaries of SolarWorld AG.[43]

The Chinese domestic cyber strategy, which has a history as long as the internet, similarly shifted in 2014–2015. Like all cybersecurity developments and related policies, the domestic cyber strategy can be linked to China's fifteen-year grand strategy. Issued by the State Council, it is more precisely entitled the National Program for the Development of Science and Technology in the Medium and Long Term 2006-2020. The grand strategy's goal includes the desire to secure domestic cyberspace in advance, using cyberattacks on foreign entities, in a preemptive defense of vulnerable civilian cyberspace. The grand strategy's goal also includes domestic innovation while becoming increasingly more integrated into international production networks.[44]

China moved forward on a determined domestic policy of enhancing its computer and information industry as a key link in advancing the new industrialization drive. The increasingly technology-based national economy and modern service industry impose a higher demand for the development of cyber technology. It includes breakthroughs in core technologies for integrated circuits and key components, major software, high-performance computers, broadband mobile telecommunication, and the next-generation internet in order to upgrade indigenous development capability and overall technological level. In addition, it focuses on creating highly credible networks and developing network information security technologies and products, as well as e-government and e-commerce platforms.[45] There has also been a focus on major next-generation internet technologies, including domestic production of high performance, dependable computers and developing security technologies concerning national infrastructure information networks and important information systems, develop novel coding technologies for network survival under complex large systems, active real-time protection, and safe storage.[46]

The current Chinese cyber strategies are based on the previous decade and a half of strategies, which were a steady buildup to this current strategy. Beginning as early as 2000, China's military cyber strategy was at the forefront when the Central Military Commission called for a study of people's war, a Chinese military theory of asymmetric warfare centered on maintaining popular support for a long and drawn out war against an enemy, under conditions of "informationalization." The Chinese cyber strategy, called Integrated Network Electronic Warfare,[47] consolidated the offensive mission for both computer network attack and electronic warfare[48] under the military General Staff Department, the highest organizational authority in the military responsible for daily administrative duties that was disbanded in a 2016 military reform shake-up when its operations were consolidated into the Joint Staff Department of the Central Military Commission. This early military cyber strategy focused on the two aspects of cyberwarfare that were well understood at the time. First, the use of cyber capabilities are central in developing intelligence, surveillance, and reconnaissance from diverse sources and assimilating these vast amounts of data in short periods of time. For instance, without accurate target locations, precision-guided weapons are ineffective. Second, disrupting an adversary's command, control, communications, and computer networks before and during operations, especially in near real time, was an important component of military cyber strategy.[49]

The guiding doctrine under this early strategy was called Local War Under Informationized Conditions.[50] This doctrine outlines the effort to develop fully networked cyber operations capable of coordinating military operations on land, in air, at sea, in space, and in cyber realms. The goal is to establish control of a rival's cyber operations and maintain dominance in the early stages of a kinetic conflict.[51] According to the US Center for Strategic and Budgetary Assessments, China appeared

> to be pursuing an overall grand strategy that has two main tiers. The upper-tier, peacetime component is to create a disposition of power so favorable to China that it will not actually have to use military force to secure its interests. The lower-tier component addresses the possibility that China might one day have to use force to secure its interests, thus concentrated increasingly on strikes against weaknesses of an adversary's information and support systems in hopes of paralyzing and collapsing the opponent in a single, stunning blow. The underlying goal is to develop plans, stratagems, and tactics that will enable the military to "win victory before the first battle." One manifestation of this strategy is the ongoing interest of the PLA theorists in developing "secret weapons that strike the enemy's most vulnerable point, at precisely the decisive moment."[52]

Then a white paper, *China's National Defense in 2004*, outlined a national strategy that asserted cyber military operations had "become the key factor in enhancing the warfighting capability of the armed forces"[53] and that the military takes cyber operations "as its orientation and strategic focus."[54] Chinese military doctrine advocates a combination of cyber and electronic warfare capabilities in the early stages of conflict.[55] At that time, before the 2015 military reorganization, both the 2004 white paper and the noted expert on the Chinese military, You Ji,[56] identified the air force as being responsible for information operations and information countermeasures. Other cyber responsibilities laid with the military then-General Staff Department.[57]

Early on, Chinese military strategists viewed cyber dominance as a key goal at the strategic and campaign level, according to the 2005 Science of Military Strategy.[58] The strategy relies on applying electronic warfare and computer network operations against an adversary's command, control, communications, computers, intelligence, surveillance, and reconnaissance (C4ISR) networks, and other essential information systems. These cyber tools should be widely employed in the earliest phases of a conflict and possibly preemptively against an enemy's information and C4ISR systems.

Additional to the core military objective, other objectives have emerged. The primary objective of the 2005 cyber strategy is to deny an enemy access to information essential for continued combat operations, ideally before other forces engage in combat.[59] For instance, according to officials and military researchers in Taiwan, China would launch an assault on Taiwan with cyberattacks aimed at crippling communications inside the island and with the United States in order to rob Taiwan's military commanders of the means to receive intelligence and pass on orders, and destroy the electronic targeting systems needed for missile defense.[60] A secondary objective is to target citizens' perception and belief systems through information deception and psychological attack.[61] For instance, the official website of a political party that China lists as an impediment to reunification with Taiwan, the Democratic

Progressive Party's (DPP), was replaced with a derogatory message—clearly a Chinese cyberattack by an unnamed official who described the party as "spreading fake news to create dissent in Taiwan society."[62] A third objective is strategic deterrence, which some Chinese military strategists see as comparable to nuclear weapons but possessing greater precision, leaving far fewer casualties, and having a longer range than most other weapons.[63] For instance, a cyberattack launched from computers in China burrowed deeply into satellite operators, defense contractors, and telecommunications companies in 2018 in order to control satellites, change the positions of the orbiting devices, and disrupt data traffic critical to phone and some internet links, as well as mapping and positioning data, thus winning a battle without firing a shot.[64]

Geng Yansheng, a spokesperson for China's Defense Ministry, was quoted as saying that the military set up the cyberwar unit—or a "cyber blue team"—to support its military training and upgrade the army's internet security defense. A report from China's state-owned Xinhua News Agency noted that Geng's comments came after the *PLA Daily*—the official military newspaper—on May 17, 2011, revealed the existence of a cyberwarfare unit. The blue team had conducted a synchronized cyber exercise with different military units in late April 2011 as part of hybrid warfare exercises,[65] originally under the now reorganized Guangzhou Military Region. Presumably, it currently works in some fashion under the Strategic Support Force.

Another early objective of cyber strategy in China—an economic cyber strategy that has been greatly modified since the 2014–2015 shift—was cyberespionage. Most countries engage in some sort of espionage of each other's governments. However, in the initial stages of China's cyber strategy from 2006 to 2014, China's military was continually active in the cyberespionage involving commercial interests as opposed to government secrets. Some scholars argue that commercial espionage was seen as necessary for building the Chinese economy.[66] A massive commercial cyberespionage campaign was conducted by advanced persistent threat group APT1/Unit 61398—a single organization of operators. Mandiant, a FireEye company, observed that since 2006, APT1/Unit 61398, compromising 141 companies in 20 major industries, has been a long-running and extensive cyberespionage commercial campaign whose capability is in large part enabled by direct government support from the military's Unit 61398.[67] As late as 2011, APT1/Unit 61398 successfully compromised at least seventeen new targets operating in ten different industries.[68] For instance, "Once the group establishes access to a victim's network, they continue to access it periodically over several months or years to steal large volumes of valuable intellectual property, including technology blueprints, proprietary manufacturing processes, test results, business plans, pricing documents, partnership agreements, emails and contact lists from victim organizations' leadership."[69] However, by 2017, APT1/Unit

61398 was mostly disbanded[70] as Chinese cyber strategy was completing its shift from volume to sophistication and from primarily commercial to primarily government objectives.

Yet another objective of cyber strategy in China—a domestic cyber strategy that has been greatly enhanced since the 2014–2015 shift—is a focus on domestic groups suspected of fostering unrest. For instance, China uses cyberoperations to keep track of perceived domestic threats, such as ethnic minorities like the Tibetans and the Uyghurs, using different technologies. One technology can track communications in the languages of the state's diverse ethnic groups. This system can monitor voice calls, text sent via the internet, and even communications embedded in images or graphics to flag up "possible social unrest."[71] This communication technology is aimed at local authorities in areas where security officials do not know the local language but can now have firsthand, real-time access to intelligence information.[72] For instance, one of the specific surveillance systems uncovered in 2009, named GhostNet, stole information from the computers used by the Dalai Lama and the Tibetan community living in India.[73]

Another set of technologies used to surveil domestic minorities involves physical tracking through DNA and facial recognition technologies. The police are now using facial recognition technology in cities like Hangzhou and Wenzhou in the east and in the central city of Sanmenxia. Almost two dozen police departments in sixteen provinces and regions sought this technology since 2018. Police departments and tech companies described the use of these technologies as "minority identification."[74] This domestic type of cyber operations has seen a boom in the post-2014–2015 strategy realignment.

While economic and domestic cyber strategies remain strong, military cyber strategy remains central in China. This is clearly indicated by improvements to the leadership and management system for services and military equipment, such as the establishment of the Strategic Support Force. In the 2019 *China's National Defense in the New Era*, China reiterates its focus on the ability of cyber technology to impact "national security, economic growth and social development."[75]

CYBERESPIONAGE

Chinese cyberespionage has been a multi-decade issue, similar to Russia and the United States. Beijing has been accused of conducting high-level cyberespionage throughout the entire era of shadow warfare. While espionage is not new to warfare, cyber or otherwise, the use of military agencies to conduct international espionage for commercial purposes is fairly unusual.

While Chinese cyberespionage has shifted some of its focus away from commercial espionage, it still uses its cyber operators to steal commercial secrets from around the world. Domestically, China's cyberespionage takes on more usual forms, like looking for possible terrorism among both its citizens and foreign visitors, as well as using cyberespionage to determine levels of social unrest. But it also looks closely at its domestic and multinational commercial entities.

In one measure of international cyberespionage, the daily barrage of cyberespionage on Western commercial entities has diminished. For instance, APT 1/Unit 61398 appears to be largely out of business, its team dispersed to other military, private, and intelligence units. There are several reasons for this change in Chinese cyberespionage. First, the change is part of President Xi Jinping's broad effort to bring the Chinese military—one of the main sponsors of commercial and governmental cyberespionage—further under his control, according to a study by the iSight intelligence unit of FireEye.[76] Second, the Chinese cyberattacks shifted focus to Russia for a while, then to South Korea and Vietnam, and occasionally aimed at targets related to disputes over claims in the South China Sea. Third, the report concludes that Chinese attacks, while decreasing in volume, have increased in sophistication, picking targets more carefully.[77] Finally, and perhaps most importantly, the current emphasis within China, after conducting economic cyberespionage on its competitors for decades, is on developing its own economic capability in a push to become a world-class player in technology development.[78]

The early cyberespionage from China was clearly described in the Mandiant report, as part of a FireEye probe, before APT1/Unit 61398 was disbanded and reassigned.[79] The analysis made clear that APT1 was a single organization of cyber operators conducting a cyberespionage campaign against a broad range of targets since at least 2006 and assessed that it was, in fact, the same entity as military Unit 61398. It was a prolific cyberespionage group. The advanced persistent threat group APT1 waged continuous and extensive cyberespionage operations because it received direct government support. The Mandiant group's investigation found that military Unit 61398 was similar to APT1 in mission, capabilities, and resources and was located in precisely the same area—four large networks in Shanghai, two of which serve Pudong—from which APT1 activity originates.

Military Unit 61398 was part of the Second Bureau of the PLA's General Staff Department's Third Department[80] before the reorganization. Unit 61398 was partially situated on Datong Road in Gaoqiaozhen, which is sited in Shanghai's Pudong district located east of the Huangpu River.[81] It was staffed by people trained in computer security, computer network operations, and English language, who number in the hundreds, if not thousands, based on the size of Unit 61398's physical infrastructure: the central building is a

130,663-square-foot facility, 12-stories high, built in early 2007. The special fiber optic communications infrastructure for the unit was provided by China Telecom in the name of national defense.[82] Nonetheless, at its height, APT1/Unit 61398 used cyberespionage to steal commercial data meant to enhance China's economic development.

In these early years, the cyberattack methodology of APT1/Unit 61398 was honed and designed to steal large volumes of valuable intellectual property. Once APT1/Unit 61398 established access, this advanced persistent threat group periodically revisited the target's network over several months or years to steal broad categories of intellectual property, including technology blueprints, proprietary manufacturing processes, test results, business plans, pricing documents, partnership agreements, emails, and contact lists from target organizations' leadership. Unit 61398 used unique cyberweapons and techniques, including two utilities designed to steal email—GETMAIL and MAPIGET. APT1/Unit 61398 maintained access to victim networks for an average of 356 days. Although the specific entity was not identified, the longest time period APT1/Unit 61398 maintained access to a victim's network was 1,764 days, or 4 years and 10 months.[83]

Near the end of an era that focused on stealing commercial information, Chinese cyberespionage also zeroed in on unusual government data. For instance, in 2015, Chinese cyberespionage compromised the data of over twenty-one million Americans at the US Office of Personnel Management. Evidence of the intentions of this operation remains nonexistent in open sources. There is no indication of this data being used anywhere and no indication of the purpose for which it was taken.[84] Additionally, Chinese cyberespionage illegally accessed dozens of computers at the Federal Deposit Insurance Corporation—an independent federal agency insuring deposits in US banks—in 2016.[85] Again, the value or purpose of this data to China remains unclear.

Domestic cyberespionage in China, part of shadow warfare in all the great powers, was established early in the Chinese use of the internet with the Great Firewall. Some instances that prompted even more censorship on the internet included the thirtieth anniversary of the 1989 Tiananmen Square protests and the 2019–2020 anti-government protests in Hong Kong. Government administrators closed individual accounts on the WeChat social media platform for minor infractions such as commenting on environmental disasters and removing tens of thousands of accounts for allegedly "harmful" content on a quarterly basis.[86] Citizens removed from WeChat were no longer able to access essential common life tasks such as transportation and banking.[87] Recently, China's domestic cyberespionage has focused with its national internet and has developed into a social credit system based on data collection, which operates in this wholly independent national internet behind the Great Firewall. This domestic data collection is part of the Chinese government's plan to "comprehensively move social credit system construction forward"[88] in 2020. The government is proposing a social credit system as a desirable way to measure and enhance "trust" nationwide and to build a culture of "sincerity." The policy states: "It will forge a public opinion environment where keeping trust is glorious. It will strengthen sincerity in government affairs, commercial sincerity, social sincerity and the construction of judicial credibility."[89]

This national data collection system on China's internet is not a single social credit system but, rather, a wide spectrum of pilot systems, some commercial and some run by local governments. Eventually, however, the National Development and Reform Commission, a powerful central body, will have vast amounts of data available in a domestic context. As laid out by China's State Council, in striving to minimize the flaws of existing data collection systems, "The government is responsible for formulating and implementing development plans, completing regulations and standards, fostering and supervising credit service markets. Focus on giving rein to the role of market mechanisms, coordinate and optimize resource allocation, encourage and muster social forces, broaden participation, move forward together, shape joint forces for social credit system construction."[90]

Moreover, the credit system wants to limit commercial swindles, sales of counterfeit products, tax evasion, and fraudulent financial claims. While not assigning a single score that will determine every aspect of every citizen's life,[91] the social credit system may not just lessen crime but may also monitor and restrict dissidents and other undesirable citizens with serious social consequences. For instance, during the 2020 COVID-19 pandemic, China integrated the social credit system into its approach to managing the spread of the pandemic. In some cases, citizens who hide their travel or medical history—and thus their potential exposure to COVID-19—can have their personal social credit scores subtracted or be put on a banned list.[92] Some localities have also incorporated activities such as the spreading of rumors and hoarding of products as social credit violations. Given that COVID-19 is speculated as arising from a cross-species transmission, the government has made the consumption of specific animals posing health risks as banned within the social credit system.[93]

Although there is no single system, there have been numerous pilots, both public and private. Of the pilot systems, 80 percent of respondents in a Chinese survey[94] approve of both commercial and government-run data collection programs. In one commercial pilot program, now ended, the government allowed private companies to trial systems and algorithms for social credit scores, including two widely covered projects: one by a partner of the social-network giant Tencent and developer of the messaging app WeChat, and another by Sesame Credit, run by the Ant Financial Services Group (AFSG), an affiliate company of Alibaba.[95] These private systems, as pilots for the social credit system, appear to have ended in 2017.

However, commercial entities still feature first and foremost in Chinese social credit systems. For instance, commercial entities retain good standing if they pay taxes on time and lose good standing for substandard or unsanitary products as determined by the data collection—a sore point for Chinese citizens due to frequent scams and food safety scandals.[96] Chinese citizens see social credit systems as a reliable source of information on the trustworthiness of commercial entities, social organizations, and possible scams to such an extent that 76 percent of people queried[97] responded that a general lack of trust in Chinese society is a problem. Respondents see social credit not as undue surveillance or excessive data collection but as a helpful means of punishing polluters, reducing substandard products, and otherwise disciplining negligent commercial entities in an era of rapid commercialization and economic growth.[98]

In addition to monitoring the trustworthiness of commercial entities, the social credit systems' data collection is meant to provide individual citizens with credit records. The more durable social credit system pilots have been primarily piloted by local governments. In these local government schemes—there were approximately forty-three cities running pilot programs in 2019[99]—criminal infractions lead to deductions from the overall individual credit score. The government asserts that the social credit system is a way to bring in those people left out of traditional credit systems, including low-income and rural households, and to have criminal records taken more seriously.[100] With the rapidness and China's economic expansion and massive urbanization—where people no longer know their neighbors well—this is seen as a method to increase the level of trust between citizens through expanding cyberspace as opposed to government overreach.

These programs to monitor commercial entities and citizens—both civil servants and private individuals—are being developed simultaneously with video surveillance systems and rapidly developing facial recognition software. China is now rivaling the West and Japan in implementing a pervasive national system of algorithmic surveillance and becoming a major distributor of surveillance equipment.[101] There are justifiable concerns that these closed-circuit television (CCTV) cameras combined with facial recognition networks can be used for nefarious purposes in China and elsewhere. While theses surveillance systems are used to view subway boarding in Shanghai, catching active shooters in the West, waking up drowsy workers in Japan, checking bus driver fitness in the UAE, and finding elders with dementia in Singapore,[102] they are also used more troublingly to track ethnic minorities.

The most well-known domestic cyberespionage and collection of information of ethnic minorities concern the Muslim Uyghurs in western China's restive Xinjiang province. Massive data collection and an analysis system use artificial intelligence to select categories of citizens, who can be targeted for additional surveillance that often leads to detention in political reeducation camps. The artificial intelligence-powered platform, which is supposed to determine who is a likely participant in terrorism, or separatist or criminal activities, is based solely or primarily on computer-generated findings. The platform is used both in police and in military settings and demonstrates the power of technology in all cyberespionage. It is able to amass vast amounts of specific personal data through manual searches, facial recognition cameras, social media, the social credit system, and even the use of cell phone apps to identify individuals—especially ethnic minorities like the Uyghurs—for criminal or "unpatriotic" social activities.[103] While the United States and Russia also conduct questionable domestic cyberespionage on their citizens, China has taken it to lengths not seen before.

With a 2020 goal to get systems in place—although the goal seems to be less of a deadline and more the end of a planning period[104]—the social credit system appears to be an electronic ecosystem made up of various stratagems that are all run in different ways by cities, government ministries, online payment providers, neighborhoods, libraries, and businesses, according to Chinese researchers who are designing the national scheme.[105] Although many of these subsystems may be interconnected by a web of information in cyberspace, it will not be a unified platform where one can type in one's ID and get a single score that will determine a citizen's life. This caricature of a system that doles out unique scores to 1.4 billion people—with around 46,000 born and around 19,000 dying each day—faces some technical and many political difficulties.[106] Politically, the Chinese government not only is trying to build trust for and between commercial entities and individual citizens but also runs a terrible risk if it loses this same trust from those same commercial entities and citizens. Even in China, domestic cyberespionage can only go so far.

Cyberespionage remains a central aspect of China's approach to cyberwarfare. International cyberespionage, in addition to being used for largely traditional espionage operations such as listening to adversaries' plans and operations, has the unusual element of using military operators to gather commercial information and intellectual property to grow China's economy—and from there, China's power. Domestic espionage is being developed on a massive scale in China within the confines of its own carefully curated national internet.

CYBERATTACKS

Cyberattacks are an important part of the Chinese shadow warfare strategy and arsenal. China's cyberattacks seem to focus on obtaining either military information like hardware design or governmental information on people and systems. As the Chinese cyber strategy shifts from a more commercial focus to a governmental and military focus, more emphasis is placed on the external operation of control systems in everything from aircraft to electric grids as potential targets of cyberattacks. The shift in targets reflects a strong emphasis on power projection and an understanding of a potential adversary's capabilities. What follows are some of the acknowledged cyberattacks by China.

China's cyberattacks against Russia escalated between 2015 and 2016.[107] The attacks appear to correlate first to the cease-fire with, and then resumption of cyberattacks against, the United States. For instance, in the aftermath of a September 2015 agreement between China's president Xi Jinping and then-US president Barack Obama promising not to engage in commercial cyberespionage,[108] the Russian Defense Ministry and the Federal Security Service were formulating measures against an increase in Chinese cyberattacks. Specifically, NetTraveler, a piece of malware linked to China, was being used against weapons manufacturers and threatened national security despite an information security agreement signed by Moscow and Beijing in May 2015. More than fifty types of Chinese Trojan viruses attacked dozens of Russian commercial entities and government institutions in 2016, including seven military enterprises specializing in missiles, radar and naval technology, five government ministries, four aviation businesses, and two commercial entities involved in the nuclear industry.[109] State-run tank manufacturer, Uralvagonzavod, and Russian Helicopters were among those attacked. These cyberattacks shifted to the United States again in 2017, following statements concerning the need for tariffs on Chinese goods made by President Donald Trump[110] and the advent of the 2018 US-China trade war. As cyberattacks against the United States re-intensified during the Trump administration, cyberattacks against Russia lessened during the same period. It appears that the strategic triangle between China, Russia, and the United States continues well into the era of shadow warfare.

In 2017, Chinese cyberattacks targeted South Korean entities involved in deploying a missile defense system.[111] After the Chinese government raised concerns regarding the deployment of the US-developed Terminal High-Altitude Air Defense (THAAD) system in South Korea, cyberattacks targeted South Korean military, government, and defense industry networks. These include a DDoS attack against the website of South Korea's Ministry of Foreign Affairs, spear-phishing emails carrying attachments loaded with malware, and downloading malware onto websites frequented by military, government, and defense industry officials.[112]

In 2013, Chinese cyberattacks targeted the Australian Department of Defence, the prime minister's and the cabinet offices, the Departments of Foreign Affairs and Trade, as well as the Reserve Bank and the Bureau of

Statistics, in sustained cyberattack operations. Even the preeminent Australian Security Intelligence Organisation—the top national security agency—was the target of a cyberattack when a contractor involved with building the new headquarters in Canberra had the blueprints stolen in a cyberattack, including the building's security and communications systems, its floor plan, and its server locations. This particular operation left the Australian spy agency vulnerable to even further cyberattacks.[113]

A Chinese cyberattack targeted Defence Research and Development Canada—a civilian agency of the Canadian Department of National Defence meant to provide the Canadian armed forces and other government departments with technology and data. The Chinese cyberattack gave access to highly classified federal information and also forced the Finance Department and Treasury Board—the federal government's two main economic nerve centers—off the internet. The cyberattack, first detected in early 2011, left the Canadian government rushing to determine how much sensitive information may have been stolen.[114]

In September 2012, Telvent Canada discovered a cyberattack by APT1/ Unit 61398 that included accessing its control systems and taking project files. Telvent Canada, now owned by Schneider Electric, designs software that gives oil and gas pipeline companies and power grid operators remote access to valves, switches, and security systems. Telvent keeps detailed blueprints on more than half of all the oil and gas pipelines in North and South America and has access to their systems. Telvent Canada cut access as soon as the intrusion was discovered, so that the attackers could not take command of the systems. The cyberattack on Telvent Canada is particularly dangerous because of the offensive capability to take out critical systems across the continent.[115]

Another cyberattack was discovered by the Communications Security Establishment, Canada's national cryptologic agency, in 2014. The target of the cyberattack was the National Research Council—Canada's largest governmental science and research organization. The National Research Council computers operate outside those of the government of Canada as a whole, so it is unlikely that the cyberattack breached other parts of the government.[116] However, the National Research Council's mission is to have an impact by advancing knowledge, applying leading-edge technologies, and working with other innovators to find creative, relevant, and sustainable solutions to Canada's current and future economic, social, and environmental challenges. The council works closely with thousands of Canadian firms. This type of attack means that proprietary information could be lost on many scientific discoveries, both for military applications and for copyright information, as well as a general projection for the direction of Canada's scientific community.

India is ranked high among countries attacked by mobile malware, malicious software that targets cell phones or other wireless devices.[117] In 2017, India claimed that a cyberattack launched by China on May 23 caused the crash of a Sukhoi 30 aircraft, part of India's Air Force fleet meant for air warfare on the India-China border. The crash killed two pilots. The wreckage of the plane was discovered three days later, and an analysis of the crash was carried out by the Indian Air Force. The inquiry determined that the flying aircraft suffered a cyberattack while it was airborne but did not come to any other definite conclusions. The US Federal Aviation Administration reported that the chances of downing a fighter jet midair using cyberweapons is possible but drew no conclusion about the Sukhoi 30 crash.[118]

The Chinese government cyberattacks on the United States go back to the early days of the internet. Titan Rain is the informal code name for a yearlong barrage of Chinese cyberattacks directed against the United States. Beginning around 2003, the US government computer networks and websites were bombarded with attacks that appeared to be coming from China, perhaps related to the US bombing of the Chinese embassy in Belgrade during the Kosovo conflict in 1999.[119] Cyberattacks, likely from China's military, targeted computer networks in the US Department of Defense and other agencies, including the Department of State, the Department of Energy, and the DHS, as well as various defense contractors.[120]

In the decade following Titian Rain, according to *The New York Times*, cyberattacks coming from China's military APT1/Unit 61398 focused not just on stealing information but obtaining the ability to manipulate critical US infrastructure like power grids and other utilities. Digital Bond, a small security firm that specializes in industrial control computers for infrastructure, was attacked in 2012 by APT1/Unit 61398 in an attempt to access confidential information about Digital Bond's clients, which include a major water project, a power plant, and a mining company. Although the attack on Digital Bond was declared ineffective, cyberattacks of these types are designed to gain command of industrial control systems.[121]

Another cyberattack by APT1/Unit 61398 was made on the National Electrical Manufacturers Association, a lobbying group that represents companies that make components for power grids. Then-president Barack Obama alluded to this concern in the 2013 State of the Union speech, saying: "Now our enemies are also seeking the ability to sabotage our power grid, our financial institutions, our air-traffic control systems. We cannot look back years from now and wonder why we did nothing."[122] Not only has the United States put malware in China's infrastructure in retaliation, but these types of moves have strained the relations between the United States and China well into the Trump administration.

Yet another cyberattack by APT1/Unit 61398 was made on a contractor for the National Geospatial-Intelligence Agency, a combat support body under the aegis of the US Department of Defense and a member of the intelligence community. The primary mission of the National Geospatial-Intelligence Agency is collecting, analyzing, and distributing geospatial intelligence in support of national security. This cyberattack, too, was rebuffed.[123]

While not a full-fledged disinformation cyberattack, Chinese foreign ministry personnel have taken to using social media—including Twitter, YouTube, and Facebook—to support a positive message about China.[124] Despite the fact that Twitter is banned inside China, government officials are using social media outside China in an effort to increase its influence. From concerns over Huawei to the detention camps in Xinjiang to apprehensions about the early handling of the COVID-19 outbreak, China is facing challenges to its global reputation. In response, China is running a concerted disinformation campaign to improve its reputation. The Twitter offensive has only been minimally effective in the first year, because unlike on Chinese domestic social media such as Weibo and WeChat, comments and challenges abound. Nonetheless, China's ambassador to the United States, Cui Tiankai, joined Twitter in June 2019; China's Foreign Ministry and Liu Xiaoming, the ambassador to Britain, followed in October 2019.[125] Venturing outside the Great Firewall and spending millions of dollars promoting their content on social media may have had limited initial success, but the slow start is unlikely to end this disinformation campaign.[126] Anyone who watched it battle over what information about China makes it onto Wikipedia for many years, a campaign in which Beijing is currently doing rather well, should recognize that China may end up as a force on international social media once it gets the hang of it.

While the main focus of Chinese cyberattacks has always been on commercial entities to harvest intellectual property and technology blueprints, major cyberattacks are also directed toward foreign government operations and military secrets. For instance, the US F-22 and F-35 jets, as well as the Russian Sukhoi Su-27, Sukhoi Su-33, and MiG-21, all appear to be copied in Chinese fighter planes. The cyberattacks also targeted foreign infrastructure, useful for sabotage or in the context of hybrid warfare. Many of the cyberattacks are either assessing the vulnerability of foreign industrial control systems or leaving behind malware in infrastructure grids for potential later use. Similarly, the Chinese government is interested in global oil and gas companies, although the companies that were breached decline to identify themselves. If a foreign oil and gas company is beginning exploration in an area of the world that a Chinese national gas company is also interested in, China might explore the computer files of that foreign company to examine its geological readings and assessments and use those same assessments to underbid the foreign competitor.[127]

CONTROLLING CYBERSPACE

The Chinese government views cyberspace as largely controllable. Around the time when Google withdrew from China in 2010—in response to a Chinese cyberattack on Google and other US tech companies that prompted Google to stop censoring internet searches in China[128]—the State Council Information Office delivered an exultant report on its work to regulate online traffic, according to a crucial Chinese contact cited by the State Department in a cable in early 2010. The person said that "in the past, a lot of officials worried that the Web could not be controlled. . . . But through the Google incident and other increased controls and surveillance, like real-name registration, they reached a conclusion: the Web is fundamentally controllable."[129]

A central aspect of controlling cyberspace is domestic control. In an attempt to enhance control over its domestic cyberspace, China adopted cybersecurity legislation to address growing threats of cyberattacks, in addition to the well-known Great Firewall, which allows the Chinese government to control its own cyberspace. The Cybersecurity Law, mentioned earlier, took effect in June 2017 and was labeled an "objective need" of China as a major cyber power.[130] The law restricts foreign technology companies from operating in sectors that are considered critical and include requirements for security reviews and for data to be stored on servers in China. This national security law intends to make all key cyber network infrastructure and systems even more secure and controllable. "China's government has come to recognize that cyberspace immediately and profoundly impacts on many if not all aspects of national security," says Rogier Creemers, a Sinologist at Leiden University. "It is a national space, it is a space for military action, for important economic action, for criminal action and for espionage."[131]

Economic success in cyberspace, like all economic development, is a high priority for China. China is leading the world in e-commerce, claiming over 40 percent of the world's 2017 transactions.[132] This makes the domestic economy vulnerable to disruptive cyberattacks. China's comprehensive e-commerce law took effect on January 1, 2019, bringing increased pressure on online retail companies in order to fight the sale of counterfeit and inferior merchandise on their platforms. The law is part of an overall effort to develop China's e-commerce market.[133] It covers the requirement for registration and licensing of e-commerce operators, taxation, electronic payment, and e-commerce dispute resolution, as well as the protection of intellectual property. China is acutely aware that economic power has been central to its rise. The protection of the economy is a top priority. Controlling cyberspace similarly focuses on military operations in and through cyberspace. In addition to using the military to assist the Chinese economy through international cyberespionage, China's military uses cyber operations to command and control its own military operations and disrupt the command and control of others. For domestic concerns, however, it appears that the MSS is in control, although perhaps working in conjunction with the Chinese military.

Perhaps the most pertinent perspective regarding controlling cyberspace is China's international stance. Within the two broad camps regarding how international cybersecurity should be achieved and organized, China emphasizes not only the sovereignty of cyberspace but also a preference for international negotiations on the norms and rules of international conduct in cyberspace.[134] The focus of China's interest in international negotiations is not only on the legality of cyberattacks and cyberwarfare but also on the dominant issue of which states control international cyberspace and the strong preference that China is one of those states. China, in its involvement in the international control of cyberspace, is using the terminology of the need for traffic rules for the information highway,[135] which draws on the language used in the Clinton administration's policy.[136]

One of the major differences between the view of the Chinese government and the view from the other camp to which the United States belongs is whether existing international law covers the new technology and strategies of cyberwarfare. China acknowledges that some international law applies to cyberwarfare, but large differences remain. Some of these differences focus on prohibitions on the use of force in the context of cyberattacks, cyberespionage, and the promulgation of disinformation campaigns. Arguing that the existing prohibition of state-on-state cyberattacks is sufficient, China insists that the other camp's additional reliance on the right to self-defense and the applicability of international humanitarian law are both wrong and imply the legitimacy of international cyberattacks and cyberespionage. The Chinese government has consistently adhered to a strict interpretation of these provisions, stating that the "use of force shall not be resorted to without the authorization of the Security Council with the exception of self-defense under armed attack."[137] That is, the Chinese government is trying to draw a hard line between cyberwarfare and kinetic warfare. Even intensive cyberwarfare—something China has honed to a fine art—should not be a reason to resort to kinetic warfare, and domestic use of cyberespionage should never be considered under international humanitarian law since this is under the exclusive rights of the sovereign state. This is not a surprising stance for China to take.

This interpretation rejects the possibility of any other legal use of force by a state, including humanitarian interventions to protect civilians from war crimes or genocide.[138] The Chinese stand on these norms and instruments of international law is strictly textualist, political, and principled. Building on the other camp's proposition that existing international law is sufficient, the Chinese government's view is that the prohibition of the use of force should be read as absolute in the context of cyberwarfare.[139] China does not want any other state to consider China's shadow war efforts as grounds for open kinetic warfare.

China argues that states have the right to control their own cyberspace, much like any other domain or territory, also known as cyber sovereignty.[140] This is remarkably similar to the Russian position on this issue, although China is closer to actually achieving it. According to the Chinese government's view, sovereignty is an absolute concept that only the sovereign state itself can condition. Thus, each country has the right to manage its own cyberspace in accordance with its domestic legislation. The Chinese government made it clear that it has a sovereign right to stop cyber traffic (both incoming and outgoing) at its borders, on the grounds that each country has the right to manage its own cyberspace in accordance with its domestic legislation. Such a view, again, is China's principled stand and within its long-standing reading of international law.[141]

The Chinese government does not consider cyberspace an unstoppable force for freedom of information.[142] China has been remarkably successful in creating a sovereign cyberspace within its own territory through the Great Firewall, and Beijing relies heavily on cyberspace for its growing economy as well as its increasingly modern military. Cyberattacks on industrial controls can be devastating to China's infrastructure. China's international stance on controlling cyberspace reflects its strong preference to control and protect its domestic cyberspace, limit cyberattacks, both domestic and international, and to continue to use cyberspace to conduct cyberespionage and to develop its economy.

SHADOW WARFARE POLICY

Fundamentally, China's shadow warfare policy is guided by political, economic, and social realities. China's chief political reality is the desire for political stability, primarily domestically but also internationally. The Chinese government worries about the Uyghurs, with its separatist movement, to the west; Taiwan, with its lean toward statehood, to the east; and Hong Kong, with its preference for democracy, to the south. If domestic control of cyberspace aids this, then the government is interested. Economically, the reality of the Chinese government is an economy that is increasingly present online with both e-commerce and business contact with distributors. The Chinese government has made clear its interest in cyberespionage to grow its economy. The social realities that underline the Chinese government's policy include both the political quest for stability and the economic quest for growth. These are two central elements to the plan for China's continued rise and success. Second, the main institutions for creating Chinese government policy are both military and civil. The Chinese military is often the international arm and instigator of aggressive cyberespionage and weapons use. The civil institutions are the primary agents of implementing domestic cyber laws and policies. All of these institutions are firmly under the control of the Chinese presidency and the politburo.

Finally, cyberwar policy in China, like modern China itself, reflects both the new and the old. The new policy is more of an approach than a philosophy reflected in China's focus on leapfrogging technologies and thus its attraction to all things cyber. China is technologically leapfrogging over the developed world in an attempt to erase centuries of technological phobia and reemerge as the world player—the one who invented gunpowder and paper. China's Big Three tech companies—known collectively as BAT (Baidu, Alibaba, Tencent)—and specialized units of the Chinese military are reaching new technological heights. Beijing has achieved the scientific and technological feats that herald its arrival as an innovation cyber power across multiple industries, from communications technology to renewable energy. For instance, China successfully tested the world's first quantum satellite communication—relying on quantum entanglement physics to exchange provably secure messages—on one end and, on the other, added, in the first half of 2017, new solar energy generation capacity equal to half of the solar bases installed in the United States in 2016.

The old is the reliance on Sunzi's *The Art of War* and has some parallels with Russia's refocusing on the traditional elements of culture. Sunzi's classic regarding the rules of war is of great influence inside and outside of China, and it is especially useful in cyberwarfare given the book's emphasis on how to fight wars without actually having to do battle. Although written 2,500 years ago, *The Art of War* applies easily to shadow warfare. Nowhere is deception more critical than in cyber defense and offense. Most of defensive and offensive cyberwar is designed to confuse adversaries into making mistakes. Sunzi has a lot to say about deception and its critical importance in warfare: "All warfare is based on deception. Hence, when able to attack, seem as if unable to attack; when using forces actively, seem inactive; when nearby, make the enemy believe you are far away; when far away, make the enemy believe you are nearby. Hold out baits to entice the enemy to act. Feign disorder and strike him when he seeks to take advantage."[143] A central element in Chinese cyberwar policy is to appear to have more when it has less and to appear to have less when it has more. Cyberwarfare policy in China combines the new with the old to reassert China onto the world stage.

The Chinese government is confident in its role as a leader in cyberwarfare. Yang Heqing, an official on the National People's Congress Standing Committee, says that cyber power is deeply linked to China's national security and development. "China is an internet power, and as one of the countries that faces the greatest internet security risks, urgently needs to establish and perfect network security legal systems."[144] The Chinese cyber approach has clearly shifted in recent years with expanding goals and increased sophistication in strategy and targets. It has also shifted from predominantly economic cyber targets to predominantly governmental and infrastructure targets. This is because China sees itself as an equal to the United States and Russia in many regards, most notably in its ability to conduct shadow warfare. Along with the United States and Russia policies on cyberwarfare, China has joined in the global battle for cyber dominance, for national influence, and for control over cyberspace to augment its kinetic war capabilities, to build a strong place in the global economy, and to maintain a dominant world presence.

Chapter 5

Cyberwar Policy

International agreements are written in rooms where long tables are arranged in a rectangular shape. There are name cards, and small bowls of candies, and drinks of water or tea. The participants take turns speaking in cordial, yet strong, voices. The table talk is broken up by opportunities where participants mingle to reiterate their points or huddle in the corners of the room making tentative deals. These tentative deals are all brought together at the end of a long day or night as country teams assemble to share information and plan the strategy for the next round. In major international agreements, like the UN Convention on the Law of the Sea, these talks can go on for years as countries parry and negotiate.

One ineffectual international agreement was announced by US President Donald Trump after a two-hour meeting with President Vladimir Putin of Russia. "Putin & I discussed forming an impenetrable Cyber Security unit so that election hacking, & many other negative things, will be guarded and safe,"[1] Trump tweeted following the talks with Putin at the G20 summit in Hamburg, Germany, in July 2017. This unusual start to an international agreement on cyberspace and cyber operations did not result in anything, certainly not a meeting of minds on cybersecurity.

One reason for the failure of a cybersecurity agreement between the United States and Russia was because it did not have the depth created in the ritualized and drawn out negotiation process. But there are other reasons for the failure as well. A major reason is that each of the great powers feels that its own approach to cyber operations is working. The US and NATO camp is confident in its ability to continue to lead the technological revolution. For instance, Antonio Missiroli, NATO's assistant secretary general for emerging security challenges, gave a speech at the CyberSec European Cyber Security Forum on March 24, 2020, emphasizing the importance of innovation and digital technologies, and stressed the preparedness of NATO's strong cyber defenses.[2] Russia, a leading member of the camp to create an international treaty to limit cyberweapons, is happy to be once again heading an international coalition against the United States, which gives Moscow a certain global prominence. China, which is in the same camp as Russia when it comes to supporting an international treaty, is happy building its economy with the aid of stolen international technology while simultaneously building its own domestic technology base that is keeping its citizenry certain that China is the center of the world. The main unifying element in Russia's and China's drive for an international agreement on cyberspace, however, may end at limiting America's strong cyberspace operations. China fears the Russian preference for disinformation campaigns, much like the Russians fear Beijing's desire to replace Moscow as Washington's chief challenger for global dominance.

Another reason for the failure to reach an international agreement on cybersecurity has to do with the time span often needed to create a major multilateral security treaty even without secrecy inherent in shadow warfare. For instance, the 1648 Peace of Westphalia gave the world the concept of state sovereignty, which cyber sovereignty harks back to. The treaty took four years to negotiate, but the agreement to create broad norms—like religious freedom—took centuries to fully realize. The transfer of the norm of sovereignty from the sovereign to the state or the citizenry was not fully realized until the American and French revolutions over a century later—a sovereignty of the people that cyberwarfare may be undermining. In another more recent instance, the Strategic Arms Limitation Talks (SALT I) began with failed starts in 1964, in 1966, and in 1967, before official talks launched in November 1969. In the case of SALT I, these were bilateral agreements rather than the larger multilateral agreement that would be needed in the case of cyberwarfare talks. After several impasses, the United States and the Soviet Union announced in May 1971 that they had reached a preliminary agreement on a partial constraint on certain strategic offensive systems and on a treaty to limit ABM systems after three years of negotiations.[3] This agreement was followed by another five years of negotiations before SALT II was written. Negotiations on major changes in warfare standardly take time—time that has not fully begun in cybersecurity talks.

Another difficulty in developing a general international agreement on shadow warfare is the changeable circumstances and technologies of cyberwarfare. Areas in which cyberattacks take place—cybercrime, hacktivism, cyberespionage, weaponized cyberattacks, and cyber disinformation campaigns—are areas where new technology, new approaches, and new social media venues keep adding to the possibilities of attacks. These areas of cyberattacks bridge both private and public arenas. While some of these areas are new, some are older strategies used in political warfare. For instance, the

British identified the value of political warfare in a World War II manual as encompassing the elements of psychological warfare, ideological warfare, morale warfare, and propaganda to be used against any enemy during armed conflict.[4] These elements become part of cyberwarfare in both disinformation campaigns and cyberespionage. And yet the primary venues for disinformation campaigns and cyberespionage are the internet and social media. It is difficult to create an international agreement when the weapons and venues are frequently shifting and expanding.

In this shifting era of warfare, it is the theories upon which the primary instrument of shadow warfare—cyberwarfare—is based that are not agreed upon. Despite the different foundations, the need for an international understanding regarding rules of shadow warfare is rapidly reaching a critical stage. The notion that cyberwarfare raises few ethical questions since it does not require boots on the ground in a foreign land is reminiscent of early arguments that nuclear weapons could be for the good of humanity.[5] In his 1953 "Atoms for Peace" speech to the UN General Assembly, US president Dwight Eisenhower professed nuclear knowledge will "help us to move out of the dark chamber of horrors into the light, to find a way by which the minds of men, the hopes of men, the souls of men everywhere, can move forward towards peace and happiness and well-being."[6] This optimistic view of the world that this nuclear technology could create did not emerge as Eisenhower hoped. Nuclear weapons proliferated to include the five permanent members of the United Nations plus India, Pakistan, North Korea, and Israel. There are thirty-one countries that use nuclear power in 2020. Neither nuclear weapons nor nuclear power created the peace and happiness that Eisenhower predicted. It seems equally unlikely that shadow warfare's persistent low-intensity state of war will improve the conduct of war or the well-being of humanity. In the current development of shadow warfare, a universal agreement or solid norms concerning the rules of cyberwarfare in war theory—similar to what was done with conventional armies and traditional warfare, and with nuclear warfare and the principle of mutually assured destruction—remains elusive.

CYBERATTACKS AND ATTACKERS

The need to put some international norms and laws in place grows when the magnitude of what cyberattacks can do is viewed as a whole. In the past three decades, cyberwar has used malware to attack the control systems of power grids, dams, centrifuges, missile launchers, and electronic election systems. It is hard to argue that the damage that cyberwar can inflict is minimal. Unless there is a clearer understanding of the global magnitude of the possible and existing potential of shadow warfare, the future will be daunting. As shadow warfare develops—and through its development creates continuous warfare and the dissolution of the sharp distinctions between allies and adversaries—fierce debate over how to regulate or develop national and international policies is the first step in creating international norms. It is creating new frames of reference for liberal democracies in terms of privacy and domestic surveillance of individuals, or the citizenry's right to know about wars engaged in on its behalf, or the growing power of the leaders of governments to engage in warfare—unseen by the public and even other arms of government—that is becoming both personal and unchecked. Shadow warfare is, however, knowable and highly relevant to the lives of citizens.

It is important to see the cyber policies of the United States, Russia, and China for what they are: attempts to project power, undermine adversaries, and gain international importance. These policies are primarily military but have social, political, and economic import. Not only do cyber operations allow great powers to track their adversaries and allies, but also grant the state the capability to track its own citizens. Not only does cyber power allow states to have complex supply chains around the globe, but it also allows them to undermine each other's economic prowess. And unlike a traditional war—like the conflicts in Afghanistan, the Gulf, or Syria—cyberwarfare is not only barely visible but also never-ending since the goal is overall power projection. In the words of Sergey Lavrov, Russia's foreign minister, "The world is changing and as has always happened in history, at some point somebody's influence and power reach their peak and then somebody [else] begins to develop still faster and more effectively."[7]

THE UNITED STATES' POLICY

US cyber policy reflects its position as an early creator of shadow warfare. The United States was not only an innovator in the development of drones for surveillance and attacks, the invention of robots to detonate bombs and provide forward vision in urban warfare, the creation of technologically driven body armor for soldiers, and the use of satellites in directing weapons systems and vehicles; it was also a pioneer in cyberwarfare. The United States was also an early user of cyberweapons and still strives to remain on the cutting edge of this field in terms of technological development. It appears to prefer large, sophisticated, and devastating cyberweapons. For instance, the United States was the first to develop, likely in partnership with Israel, sophisticated weapons to attack electronic control systems.

Aggressive use of advanced cyberweapons appears to be a hallmark of US cyber policy—hence the need to stay on the cutting edge of technology. The underlying policy can be described as "the best defense is a good offense," to paraphrase George Washington.[8] Evermore sophisticated cyberweapons are manufactured with thousands of man-hours and the best minds in the US intelligence agencies. What the United States sees as the success of its cyberwar policy is reflected in its stance on the potential need for an international agreement on cyberwarfare.

So far, the US government has resisted attempts from the Russia and China camp to create an international agreement on cyberweapons or cyberspace. Relying on its role as a leader at the forefront of the expertise used for cyberattacks of all types, the United States is concerned that an international agreement—especially one led by Russia and China—is meant to limit cyberweapon technology. Moreover, Washington acts as if it has control over cyberspace, so here too American policymakers see this as a US advantage. Thus, to negotiate an international agreement is to risk curtailing the US advantages in order to either allow other major cyber powers to catch up or allow Russia and China to use their own strengths and advantages in cyberwarfare. Finally, the United States does not want an international agreement at this time because cyberwarfare continues to expand to fill numerous security needs and desires, as illustrated in chapter 2.

Like most serious cyber powers, the United States has continuously modified and enhanced its cyberwar institutions. While there are national agencies that are concerned with protecting governmental operations, the main force of US cyber operations lies with military and intelligence agencies. The main locus of power, however, remains with the president himself. This was the case during the Clinton administration, but has only become more focused at the level of the president during the Bush, Obama, and Trump administrations. While this consolidation of decision making under the top executive is not unique among the great cyber powers, it is relatively unique compared to other decision mechanisms in the United States. While it bears similarities to the US president having exclusive control over the nuclear launch button, in the case of cyberwarfare the president also has considerable control over policy, which is less the case with nuclear policy. The lack of oversight on many aspects of cyber operations can be troublesome in a liberal democracy.

Problematically for a liberal democracy, as the Edward Snowden releases clearly illustrated, the US government uses cyberespionage in its domestic environment and has for some time. Domestic cyberespionage is accelerating in the United States as elsewhere. For instance, amid the COVID-19 pandemic, the US government, through the Centers for Disease Control and Prevention—the leading US public health institute that is a federal agency under the Department of Health and Human Services and thus in the executive branch and the president's line of command—is using location data from millions of cell phones on the presence and movement of American citizens in areas of geographic interest.[9] Although prompted by the pandemic, there are troubling long-term concerns about the loss of privacy in a liberal democracy similar to the ongoing repercussions of the Patriot Act's allowance of surveillance after the 9/11 attacks amid fears of terrorism. Pandemic surveillance has been in the press but has received little congressional oversight or public debate to date.

RUSSIA'S POLICY

The Russian cyber policy reflects its foundations in combining traditional Russian disinformation with modern cyber technology. This is clearly a tactic that Russia has used, not only domestically but also in its policy toward the liberal democracies in the United States and Europe, and in the countries around its borders. It also reflects the combined eagerness of the Russian government to engage in shadow warfare, and especially in cyberespionage and disinformation campaigns, following a slow start in developing sophisticated cyberweapons in the aftermath of the collapse of the Soviet Union. While the United States was ready to go in the early 1990s, Russia took about fifteen years to catch up on the art of shadow warfare. But in most ways, Russia has caught up.[10]

Russia has built specific institutions to engage in cyberwarfare and policy centers for interfering in an adversary's society, usually by exaggerating existing social divides. Like the United States, Russia's cyber strategy is aggressive. Unlike the United States, Russia's strategy is less about cyberweapons and more about enhancing its influence by showing the weaknesses of its adversaries by exploiting social fractures through disseminating disinformation. This strategy is meant to deter those who would ignore or belittle Russia, weaken adversaries in preparation for kinetic or all-out cyberwarfare, and hopefully raise morale at home by showing that Russian leadership is as good or better than that in liberal democracies.

Russia is as good at domestic cyberespionage as its great-power rivals. In addition to the existing use of the SORM boxes—operative search measures—that monitor emails, internet usage, cell phones, Skype, text messages, and social media,[11] domestic surveillance and disinformation campaigns are ramping up. For instance, we are seeing a massive disinformation campaign using both state and social media regarding the COVID-19 pandemic.[12] According to one EU document, Russia's state-controlled media, like RT and Sputnik, are using the ambitious disinformation campaign in order to sow the seeds of panic and distrust in the United States and NATO countries.[13] Even more importantly, Russia is establishing individual citizen tracking in the face of the pandemic similar to the United States and China. One element in this tracking is a massive facial recognition system, which initially prompted an unusual public backlash, with privacy advocates filing lawsuits over unlawful surveillance, but the resistance has subsided during the COVID-19 crisis. The tracking also analyzes the social networks of those who have or are suspected of having the coronavirus. Finally, there is the use of geolocation to track coronavirus carriers: information gathered under the tracking system will be used to send texts to those who have come into contact with a coronavirus carrier and to notify regional authorities so they can place individuals in quarantine.[14] Again, there are concerns that domestic surveillance created under pandemic conditions will not cease once the health crisis is over.

The confidence engendered from a lack of criticism or oversight on domestic cyber policy and domestic cyberespionage leads to even further reaches in cyberspace. Similar to the United States and China, Russia argues that it can control cyberspace. However, unlike the United States but like China, Russia argues that cyberspace should be considered another form of territory—like land, air, and water—within a sovereign state. Moreover, Russia is a major proponent that this perception of cyberspace should be codified by an international treaty that will lead to more global peace and stability. Nonetheless, Russia rivals the United States and China in terms of being one of the most active violators of using cyber operations to destabilize global peace. This is comparable to other examples of mutual mistrust, like on nuclear disarmament, that has caused the great powers to use new technology to make the world a less stable place.

A further attempt to control or modify cyberspace is the effort to create new international norms. In a stance that has aspects similar to both the United States and China approaches, the Russian government has initially suggested giving the International Telecommunications Union (ITU)—a UN body where each government has an equal vote[15]—control over cyberspace. This has elements of the US stance that there are laws, norms, and institutions already in existence that can deal with most aspects of cyberspace and cyber-operations. Presently, internet infrastructure is governed by the ICANN—a nonprofit corporation that is responsible for allocating IP addresses and managing the domain name system which, it argues, is composed of volunteers on its board from across the world in a "bottom-up," open, and transparent process.[16] Based in the United States, with many Americans on its board, ICANN is now directly accountable to the global community of internet users.[17] Russia is not supportive of a US-dominated cyberspace and the lack of success over moving controls either to an international body like the ITU or just out of the United States. Moscow has physically moved closer to the Chinese position on cyber sovereignty, as illustrated by attempts to close down the internet inside its national boundaries in the case of an emergency. This has elements of China's stance that prefer a new international treaty on cyberspace, now including the concept of cyber sovereignty in the agreement.

CHINA'S POLICY

China's cyber policy reflects its reliance on strategy over technology, on stealth over disinformation. Nonetheless, the use of technology is on the rise as its strategy shifts and the capabilities of its adversaries grow. The main thrust of China's cyber strategy, especially in the first two decades of the era of shadow warfare, was in further building its economic strength through military cyberespionage. Now that China ranks as a top economy in the world, the thrust has shifted somewhat to domestic development of its own technology, whether it is 5G or a global positioning system.

The use of technology in cyberwarfare is on the rise as its strategy changes. For instance, the locus of cyberespionage has shifted to China Telecom—a large state-controlled telecommunications company—that has ten strategically placed internet points of presence (PoPs) across the internet mainstay of North America as well as PoPs in Europe and Asia. Vast intelligence rewards can be reaped from the hijacking, diverting, and then copying of information-rich traffic going into or crossing the United States and Canada—often unnoticed and then delivered with only small delays. This hijacking allows China Telecom to employ its distributed PoPs in the telecommunications systems of liberal democracies to selectively redirect internet traffic through China. These Chinese PoPs allow one to simply divert and copy data by controlling key transit nodes buried in a country's infrastructure.[18] Russia is alleged to use a similar tactic in cyberespionage, and it is safe to presume the United States is also involved.[19]

Disinformation is now also included in Beijing's arsenal. For instance, China launched a forceful information campaign aimed at creating a positive global discussion on its handling of the novel coronavirus outbreak. This campaign is meant to give China credit for its handling of the pandemic while highlighting other governments' missteps.[20] Building on its already vast domestic espionage program, China began a massive new surveillance project by installing software on the citizens' cell phones.[21] According to an official China news outlet *Xinhua*, the health-code app is an electronic certificate that tracks one's movements and symptoms, indicating if someone needs to self-isolate or can continue to move about.[22] The longer-term concern, however, is whether this extensive surveillance will be dialed down with the fading of the SARS-CoV-2 virus or become a more permanent feature of life in China, much like it already is for some ethnic minority groups as discussed in chapter 4.

For China, similar but not identical to Russia, a central aspect of controlling cyberspace is domestic control or cyber sovereignty. Specifically, China is largely restricting the information that is available to citizens via the internet and social media and is using the internet domestically to surveil citizens and entities. Within a controlled cyberspace, China remains interested in economic benefits. Economic success in cyberspace—China is leading the world in e-commerce—like all economic development, remains a significant concern.[23] Although the military plays a role in protecting the domestic economy, military modernization means that now China uses cyberspace for command and control of its own military operations. Of course, the military also strives to disrupt the command and control of adversaries as well.

In addition to cyber sovereignty to control cyberspace, China also advocates for international negotiations on the norms and rules of international conduct in cyberspace.[24] In part, this is a preference for predictability in a rapidly changing environment. In part, China has concerns about how the United States and other liberal democracies will interpret and use existing international law—which the liberal democracies have formulated over the centuries—to conduct cyberwarfare to China's disadvantage. Importantly, China does not want any states to consider its shadow war efforts as grounds for open kinetic warfare, which the United States has already signaled that it might. Since China is not alone either in wanting cyber sovereignty or in advocating an international agreement on cyber norms, it is becoming increasingly confident that cyber sovereignty and an international agreement on cyber norms may be the future of shadow warfare.[25]

UNCERTAIN CIRCUMSTANCES AND DIVERGENT POLICIES

The divergent policies, policy aims, and the national character of the great powers make it difficult to create a single unifying set of laws and norms in the uncertain environment of constantly evolving cyberweapons. Nonetheless, there are a few areas of tacit agreement among the three great powers. First, cyberwarfare is persistent and permanent. It may pause or shift focus, but it does not cease. Second, the distinctions between allies and adversaries are blurring, especially as cyberespionage merges into cyberattacks and allies are spied upon as are adversaries. Additionally, cyberattacks can be on undeclared actors that are neither allies nor adversaries, like the US targeting Pakistan, or China targeting South Korea, or Russia targeting Estonia. There may be a good reason to curtail these great-power behaviors, but they all act as if these are the norms. Still, there is no written or negotiated agreement.

The debate, then, is whether there can be a meeting of the minds on the specifics of the acceptable conduct of cyberwarfare that can be codified into an agreement. While there have been numerous small attempts to monitor and establish norms of cyberwarfare, actual cyber operations are far exceeding any attempt at the kind of confidence building that proceeds an international agreement. Through actual practice, major players like the United States, Russia, and China, as well as smaller but sophisticated players like Israel, the UK, and North Korea, are creating practices that may not be desirable in hindsight. If the great powers do not address agreed norms for cyberwarfare—and they have not over the past three decades—then the world may find itself unmoored and uncertain.

Several comparisons have been made in this book between Cold War-era nuclear weapons and shadow warfare cyber policy. The International Atomic Energy Agency (IAEA), a multilateral body that seeks to inhibit the use of nuclear power for military purposes, was established in 1957, just twelve years after the nuclear age began. The Nuclear Non-Proliferation treaty (NPT) entered into force in 1970, just two and a half decades after the first nuclear weapon was dropped. Unfortunately, there are no international organizations with the stature and effectiveness of the IAEA or the NPT to oversee, regulate, and attempt to lessen cyberwarfare. A few UN organizations and some regional and national forums have discussed the future of internet governance, but there has been a nominal forward movement in light of what the great powers view as their successes in using cyberwarfare and the lack of a unifying theory like mutually assured destruction.

Cyberwarfare governance currently seems out of reach. As the Council on Foreign Relations' *Defending an Open, Global, Secure, and Resilient Internet* from 2013 notes, addressing the challenges of cyberspace expresses concern over the ongoing national emphasis of cyberwarfare and that "the effects of domestic decisions spread far beyond national borders and will affect not only users, companies, nongovernmental organizations and policymakers in other countries, but also the health, stability, resilience and integrity of the global internet."[26] In addition to the strong national decisions being taken by the great powers regarding cyberwarfare, international approaches to cybersecurity have challenges as well. For instance, there will likely need to be a balancing of the movement of trade with a global regulatory framework and protection of intellectual property in a world flourishing with economic cyberespionage. There will likely need to be a new vision of national security that includes protecting critical infrastructure and domestic rights and privacy in a world with large-scale video surveillance, facial recognition software, and cell phone tracking.[27]

The world is lacking even an agreement on a definition of cyberwarfare. For instance, the Shanghai Cooperation Organization, whose members include China and Russia, defined cyberwar with a focus on the dissemination of information "harmful to the spiritual, moral and cultural spheres of other states,"[28] first in 2011 and again in the revised version of January 2015. By contrast, the United States and other major liberal democracies'[29] definition of cyberwar focuses on physical and economic damage and injury and limits political concerns as the prerogative of freedom of speech.[30] With definitions being at odds, it is harder to move toward agreement of limiting cyberwarfare.

Attempts have been made, nonetheless, to move toward some common understanding. The East-West Institute released the first joint Russian-American report in early 2011, aimed at defining the rules and the norms for cyberwarfare.[31] Prepared by a team of experts from Russia and the United States, "Working towards Rules for Governing Cyber Conflict: Rendering the Geneva and Hague Conventions in Cyberspace" explores how to extend the existing international principles that govern war to cyberspace. Ultimately, the group debated five crucial questions. In summary, the questions raised included whether in all warfare—cyberwarfare, nuclear war, or conventional war—some targets should be off limits to attacks and whether it was feasible to use special markers to designate protected zones in cyberspace. Another major issue considered in this early study is whether some cyberweapons are already banned under the Geneva Protocol.[32] This early attempt to devise rules has been followed by other attempts to achieve understanding, which saw the great powers taking stances that solidified into the two main camps facing each other today. One camp, which includes Russia and China, argues that only a treaty process can create laws and norms for cyberwarfare. Specifically, this camp argues that the norms of cyberwarfare should include sovereign control of cyberspace. The other camp, which includes the United States and most NATO countries, argues that cyberwarfare can be dealt with under existing international law, with a few modifications.

In 2013, the UN Group of Governmental Experts concluded that both the UN Charter and international law are fully applicable to state behavior in cyberspace, a position which is supported by the US camp. Moreover, the US camp's position assumes the use of the already existing institutional framework facilitated by ICANN. This is at odds with the Russia and China camp that asserts each government's sovereign right to regulate the internet, the main premise of cyber sovereignty—the concept that cyberspace, like territorial waters or sovereign land, is owned and regulated entirely by the country it exists within—unless it is modified by specifically agreed international commitments.[33] While the 2015 agreement between the United States and China concerning economic cyberespionage is the first accord on the issue, it has been viewed with both optimism and skepticism.[34] China's preference for a new international agreement on cyberwarfare is stated in its 2016 National Cyberspace Security Strategy:

> Support the United Nations to play a leading role in promoting the development of universally accepted international rules on cyberspace, cyberspace international counter-terrorism conventions, sound judicial assistance mechanisms

> against cybercrime, deepening policy and law, technological innovation, standards and norms, emergency response, and critical information infrastructure International cooperation in areas such as protection.[35]

The United States and NATO camp led a group specializing in international law in creating a document in 2013 that attempts to apply existing international law to cyberwarfare, the *Tallinn Manual on the International Law Applicable to Cyber Warfare*. The document is also known simply as the Tallinn Manual, named after the capital of Estonia, where it was devised. The drafting of Tallinn 1.0 failed to include states such as China and Russia. There is now a second edition, known as the Tallinn Manual 2.0, which has some input from China but none from Russia. Since there is no new international law that directly refers to cyberwarfare, perhaps the existing law of war can be seen as encompassing shadow warfare by expanding the applicability. In 2012, NATO's Cooperative Cyber Defence Centre of Excellence (CDCOE) met as a panel of international legal experts from 2009 until 2012 to go through this existing law in order to apply it to cyberwarfare. This formed the basis of the Tallinn Manual and the analysis and rules it contains.[36]

Through these rules, the manual attempts to define some of the basics of cyberwarfare. At the most fundamental level, rules of war can be reinterpreted that a cyberattack on a state can, in certain circumstances, be the equivalent of a kinetic attack. The manual also lays out that such a cyberattack is against international law, and that a state attacked in such a way has the right to retaliate. The manual also argues that certain cyberattacks, such as targeting civilians and crippling civilian infrastructure, are against the rules of war whether the instruments used are tanks or cyberweapons. Many of these laws of war are well understood in the context of traditional kinetic warfare. The manual, however, is stating that these laws of war apply to cyberwar as well. These rules are of particular interest to the United States and other liberal democracies that form the camp that insists that a large new treaty on cyberspace is not needed.[37]

China has been a bit more outspoken on the issues delineated in the manual. One China scholar argued that the manual set the factors relevant to evaluating when a cyber operation rises to a use of force—severity, directness, and invasiveness—are too pliable and too low. Moreover, a state does not have the right to invoke self-defense against attacks by non-state actors, nor does a state have the right of self-defense against an imminent attack. This perspective is consistent with a Chinese interest in actively engaging in robust cyber operations without wanting to trigger a military conflict.[38] The disagreement over the Manual, however, does highlight the major cyber powers remain on different conceptual pages on how to proceed, leaving products such as the Tallinn Manuals without universal support.

Both camps in cyberwarfare theory—those both for and against a new treaty or series of agreements that define cyberspace and its rules—have some commonalities in the conduct of cyberwarfare. In addition to the areas of tacit agreement, these commonalities center upon protecting themselves against attacks and the supposed anonymity of those attacks. The quality of their cryptanalytic programs allows them to decipher who is creating a specific set of cyberespionage operations or cyberweapons with a high degree of accuracy so that while a cyber operation may be silent, it is rarely anonymous. Other commonalities include the pervasive use of cyberespionage. All believe that cyberspace is controllable. All have conducted cyberattacks. Both the United States and Russia use cyberattacks as part of a hybrid war effort, while China is using hybrid warfare in its military planning. Although the United States and China used to make significant use of privateers as part of their cyber strategy, Russia still does. These commonalities are not agreements, however.

Most important, however, is that all three major players accept a state of perpetual war through nonwar warfare as a natural condition. Perpetual war does not necessarily denote a perpetual adversary since allies and adversaries are less well defined in shadow warfare. It does, however, fundamentally change the notion of who and what makes an ally or an adversary. The extraordinary variations that shadow war allows means that some allies are also partial adversaries, and adversaries can occasionally be partial allies in the swiftly shifting sands of nonwar warfare. In shadow warfare and cyberwar, new power and new weapons present ethical challenges. The major powers are meeting goals and objectives with their cyberwar strategies and attacks. There is pressure to create new laws to limit cyberwarfare or at least to apply existing rules of law to cyberwarfare, but so far, these pressures have failed to move the major powers. They are getting what they want without them. It does not, however, leave the world as a safer place.

There are some portends about what this world of shadow warfare is becoming. For one, it is not a peaceful world regardless of what the great powers may say. It is a world filled with constant cyberwarfare. For another, it is also not a stable world. Joining with the great powers, states like Israel, Iran, North Korea, and the UK are also using shadow warfare strategies to increase their own power, inevitably destabilizing the power structure. Most troubling, again for those living in liberal democracies and for the citizens of most countries, is the loss of certain accepted democratic norms. One major norm is the balance between privacy and security. Cyber technology has made domestic espionage so simple, meaning that citizens are being surveilled at an increasingly alarming rate without much legislative oversight. Not only are all forms of electronic communications intercepted, but also CCTV cameras and facial recognition software track citizens as they go about their daily lives. GPS tracking systems, too, are now aware of citizens' coming and goings as well as their buying histories.

Another diminishing democratic norm is the domestic balance of power. In a world of shadow warfare, leaders—presidents, prime ministers, and chancellors—have access to cyberweapons over which the legislature does not have sufficient oversight, and the judiciary has not made sufficient legal rulings. In short, the world of shadow warfare helps create a norm in which individual leaders in great powers have an extraordinary capacity to inflict damage on each other without the restraints of the rules of war, without the visibility of war, and without the consent of the citizenry. If the world continues to unquestionably accept this progress toward this deepening of cyberwarfare both in the daily experience and in the wider political arena, the centuries of progress made in curbing the power of the state may be slowly chipped away, leading back to a time when princes played games with the lives of their resistless subjects.

Notes

CHAPTER 1

1. David Petraeus, "Cyber Changed War, But the Causes and Conduct of Conflict Remain Human," *The World Post*, March 29, 2017, https://www.belfercenter.org/publication/cyber-changed-war-causes-and-conduct-conflict-remain-human.

2. Thomas Rid, "Cyber War and Peace: Hacking Can Reduce Real World Violence," *Foreign Affairs*, November/December 2013, 77.

3. René Girard, *Battling to the End: Conversations With Benoît Chantre* (East Lansing, MI: Michigan State University Press, 2009).

4. Steve Coll, "The Rewards (and Risks) of Cyber War," *The New Yorker*, June 6, 2012, http://www.newyorker.com/news/daily-comment/the-rewards-and-risks-of-cyber-war.

5. Paul Robinson (ed.), *Just War in Comparative Perspective* (Abingdon-on-Thames, UK: Routledge, 2003).

6. Richard A. Clarke and Robert K. Knake, *Cyber War* (New York: HarperCollins Books, 2010).

7. James C. Mulvenon and Gregory J. Rattray (eds.), *Addressing Cyber Instability* (Cambridge, MA: Cyber Conflict Studies Association, 2012), xi.

8. Ibid., xiii.

9. David Gilbert, "Cyber War—Just the Beginning of a New Military Era," *International Business Times*, April 26, 2013, http://www.ibtimes.co.uk/articles/461688/20130426/cyber-war-beginning-new-military-era.htm.

10. Jason Andress and Steve Winterfeld, *Cyber Warfare: Techniques, Tactics and Tools for Security Practitioners* (Waltham, MA: Syngress, 2011).

11. Peter Crail, "IAEA: Syria Tried to Build Nuclear Reactor," *Arms Control Association*, March 2009, https://www.armscontrol.org/act/2009-03/iaea-syrian-reactor-explanation-suspect.

12. Martin Libicki, *Cyberdeterrence and Cyberwar* (Santa Monica, CA: Rand Publishing, 2009), xiv.

13. Joseph S. Nye, Jr., *The Future of Power in the 21st Century* (New York: Public Affairs Press, 2011).

14. Declan Walsh and Ihsanullah Tipu Mehsud, "Civilian Deaths in Drone Strikes Cited in Report," *The New York Times*, October 22, 2013, http://www.nytimes.com/2013/10/22/world/asia/civilian-deaths-in-drone-strikes-cited-in-report.html?hpw_r=0.

15. Mulvenon and Rattray (eds.), *Addressing Cyber Instability*, supra., viii.

16. Gregory J. Rattray, Chris Evans, and Jason Healey, "American Security in the Cyber Commons," in *Contested Commons: The Future of American Power in a Multipolar World*, ed. Abraham M. Denmark and Dr. James Mulvenon (Washington, DC: Center for a New American Security, 2010), 137–176.

17. Deputy Secretary of Defense Memorandum, Subject: *The Definition of Cyberspace*, Washington, D.C., May 12, 2008.

18. Mulvenon and Rattray (eds.), *Addressing Cyber Instability*, supra., ix–x.

19. Some experts dislike the use of the term "cyberwarfare" and prefer the more precise terminology of "international acts of cyber conflict," such as Jeffery Carr in *Inside Cyber Warfare: Mapping the Cyber Underworld* (Sebastopol, CA: O'Reilly Media, 2009): xiii.

20. Carr, *Inside Cyber Warfare: Mapping the Cyber Underworld*, 2nd Edition (Boston, MA: O'Reilly Media, Inc., 2012)

21. C. Gray, *Another Bloody Century—Future Warfare* (London: Weidenfeld/Nicolson, 2005), 37.

22. Sun Tzu, *On the Art of War* (2010), http://www.chinapage.com/sunzi-ehtml.

23. Thomas Rid, "Cyber War and Peace," supra., 78.

24. Richard Stiennon, *Surviving Cyberwar*, vii.

25. James A. Lewis and Katrina Timlin, *Cybersecurity and Cyberwarfare: Preliminary Assessment of National Doctrine and Organization*, Center for Strategic and International Studies, United Nations Institute for Disarmament Research, 2011.

26. Office of the US President, "Cyber Space Policy Review: Assuring a Trusted and Resilient Information and Communications Infrastructure" (Washington, D.C., 2009).

27. "N. Korea's cyber warfare unit in spotlight after attack on S. Korean bank," *Yonhap News Agency*, May 3, 2011, http://english.yonhapnews.co.kr/national/2011/05/03/78/0301000000AEN20110503010600315F.HTML.

28. Martin C. Libicki, *Cyberdeterrence and Cyberwar*, supra.

29. William Kaufmann, "The Evolution of Deterrence, 1945–1958," unpublished RAND research (1958), cited in Libicki, *Cyberdeterrence and Cyberwar*, supra, 7.

30. James C. Mulvenon and Gregory J. Rattray (eds.), *Addressing Cyber Instability*, supra., 28.

31. William A. Owens, Kenneth W. Dam, and Herbert S. Lin (eds.), *Technology, Policy, Law, and Ethics Regarding US Acquisition and Use of Cyberattack Capabilities* (Washington, D.C.: National Academies Press, 2009), 1.

32. Jeffrey Carr, *Inside Cyber Warfare: Mapping the Cyber Underworld*, 2012, supra.

33. Martin Libicki, *Cyberdeterrence and Cyberwar*, supra.

34. Jonathan Masters, "Confronting the Cyber Threat," Council on Foreign Relations Publication, March 17, 2011.

35. Rid, "Cyber War and Peace," supra., 79.

36. David E. Sanger, "Obama Order Sped Up Wave of Cyberattacks Against Iran," *The New York Times*, June 1, 2012, http://www.nytimes.com/2012/06/01/world/middleeast/obama-ordered-wave-of-cyberattacks-against-iran.html?pagewanted=all_r=0.

37. Ibid.

38. Ibid.

39. Ibid.

40. Ibid.

41. Ibid.

42. Ibid.

43. Ibid.

44. David Gilbert, "Cyber War—Just the Beginning of a New Military Era," supra.

45. Seymour M. Hersh, "The Online Threat: Should we be Worried about a Cyber War?" *The New Yorker*, November 1, 2010, http://www.newyorker.com/reporting/2010/11/01/101101fa_fact_hersh#ixzz1LP462Ulr.

46. Dorothy Denning, *Information Warfare and Security* (New York: Addison-Wesley Professional, 1998).

47. "Natural Disasters Cost US a Record $306 Billion Last Year," *CBS*, January 8, 2018, https://www.cbsnews.com/news/us-record-306-billion-natural-disasters-last-year-hurricanes-wilidfires/.

48. Josh Fruhlinger, "Top Cybersecurity Facts, Figures and Statistics for 2020," *CSO Cyber Security Report*, March 9, 2020, https://www.csoonline.com/article/3153707/security/top-5-cybersecurity-facts-figures-and-statistics.html.

49. Steve Morgan, "Top 5 Cybersecurity Facts, Figures and Statistics for 2018," *CSO Cyber Security Report*, January 23, 2018, https://www.csoonline.com/article/3153707/security/top-5-cybersecurity-facts-figures-and-statistics.html.

50. James A. Lewis, *Assessing the Risks of Cyber Terrorism, Cyber War and Other Cyber Threats* (Washington, DC: Center for Strategic and International Studies, 2002), 10.

51. David E. Sanger, David D. Kirkpatrick, and Nicole Perlroth, "The World Once Laughed at North Korean Cyberpower. No More," *The New York Times*, October 15, 2017, https://www.nytimes.com/2017/10/15/world/asia/north-korea-hacking-cyber-sony.html.

52. Ken Dilanian, "Watch Out. North Korea Keeps Getting Better at Hacking," ABC News, February 20, 2018, https://www.nbcnews.com/news/north-korea/watch-out-north-korea-keeps-getting-better-hacking-n849381.

53. Dai Davis, "Hacktivism: Good or Evil?" *ComputerWeekly.com*. http://www.computerweekly.com/opinion/Hacktivism-Good-or-Evil.

54. Ibid.

55. Ibid.

56. Ibid.

57. "Zimbabwe's Begging Bowl: Bailing out Bandits," *The Economist*, July 9, 2016, https://www.economist.com/middle-east-and-africa/2016/07/09/bailing-out-bandits.

58. Davis, "Hacktivism: Good or Evil?" supra.

59. Ibid.

60. Mark Clayton, "More Telltale Signs of Cyber Spying and Cyberattacks Arise in Middle East," *The Christian Science Monitor*, August 21, 2012, http://www.csmonitor.com/USA/2012/0821/More-telltale-signs-of-cyber-spying-and-cyber-attacks-arise-in-Middle-East-video.

61. Andy Greenberg, "New Group of Iranian Hackers Linked to Destructive Malware," *Wired*, September 20, 2017, https://www.wired.com/story/iran-hackers-apt33/

62. US Department of Justice, "US Charges Russian FSB Officers and Their Criminal Conspirators for Hacking Yahoo and Millions of Email Accounts," DOJ Press Release, March 15, 2017, https://www.justice.gov/opa/pr/us-charges-russian-fsb-officers-and-their-criminal-conspirators-hacking-yahoo-and-millions.

62. FSB Officers Protected, Directed, Facilitated and Paid Criminal Hackers.

63. Charley Snyder and Michael Sulmeyer, "The Department of Justice Makes the Next Move in the U.S.-Russia Espionage Drama," *Lawfare*, March 16, 2017, https://www.lawfareblog.com/department-justice-makes-next-move-us-russia-espionage-drama.

64. United Nations Office of Drugs and Crime, *The Use of the Internet for Terrorist Purposes* (New York, NY: United Nations, 2012).

65. Michael Sulmeyer, "Cybersecurity in the 2017 National Security Strategy," *Lawfare*, December 19, 2017, https://www.lawfareblog.com/cybersecurity-2017-national-security-strategy.

66. Lewis, *Assessing the Risks of Cyber Terrorism*, supra, 8.

67. Ibid., 1.

68. Ibid., 3.

69. Ibid., 4–5.

70. Rid, "Cyber War and Peace," supra., 82.

71. David E. Sanger, "With Spy Charges, U.S. Draws a Line That Few Others Recognize," *The New York Times*, May 19, 2014, http://www.nytimes.com/2014/05/20/us/us-treads-fine-line-in-fighting-chinese-espionage.html?hp_r=0.

72. Tabassum Zakaria, "U.S. Blames China, Russia for Cyber Espionage," *Reuters*, November 3, 2011, http://www.reuters.com/article/2011/11/03/us-usa-cyber-china-idUSTRE7A23FX20111103.

73. US National Counterintelligence Executive, *Foreign Spies Stealing US Economic Secrets in Cyberspace*, October 2011, http://www.ncix.gov/publications/reports/fecie_all/Foreign_Economic_Collection_2011.pdf.

74. Zakaria, "U.S. blames China, Russia for Cyber Espionage," supra.

75. Deutsche Welle, "Merkel Testifies on NSA Spying Affair," *Deutsche Welle*, February 16, 2017, http://www.dw.com/en/merkel-testifies-on-nsa-spying-affair/a-37576690.

76. Ibid.

77. Stephen Castle, "Report of U.S. Spying Angers European Allies," *The New York Times*, June 30, 2013, https://www.nytimes.com/2013/07/01/world/europe/europeans-angered-by-report-of-us-spying.html.

78. Scott Shane, Matthew Rosenberg, and Andrew W. Lehren, "WikiLeaks Releases Trove of Alleged C.I.A. Hacking Documents," *The New York Times*, March 7, 2017, https://www.nytimes.com/2017/03/07/world/europe/wikileaks-cia-hacking.html

79. Welle, "Merkel Testifies on NSA Spying Affair," supra.

80. Ellen Nakashima, "Newly Identified Computer Virus, used for Spying, is 20 Times Size of Stuxnet," *Washington Post*, May 28, 2012, http://www.washingtonpost.com/world/national-security/newly-identified-computer-virus-used-for-spying-is-20-times-size-of-stuxnet/2012/05/28/gJQAWa3VxU_story.html.

81. Ellen Nakashima, Greg Miller, and Julie Tate, "U.S., Israel Developed Flame Computer Virus to Slow Iranian Nuclear Efforts, Officials Say," *The Washington Post*, June 19, 2012, http://www.washingtonpost.com/world/national-security/us-israel-developed-computer-virus-to-slow-iranian-nuclear-efforts-officials-say/2012/06/19/gJQA6xBPoV_story.html.

82. Rouven Cohen, "New Massive Cyber-Attack an 'Industrial Vacuum Cleaner for Sensitive Information,'" *Forbes*, May 28, 2012.

83. Nakashima, Miller, and Tate, "U.S., Israel Developed Flame Computer Virus," supra.

84. Ibid.

85. Or could it have been that Flame came to light after the UN's telecoms body asked for help with identifying a virus found stealing data from many PCs in the Middle East? "Flame Malware Makers Send 'Suicide' Code," *BBC News*, June 2012, http://www.bbc.co.uk/news/technology-18365844.

86. Nakashima, Miller, and Tate, "U.S., Israel Developed Flame Computer Virus," supra.

87. Nakashima, "Newly Identified Computer Virus, Used for Spying," supra.

88. "Flame Malware Makers," supra.

89. Nakashima, Miller, and Tate, "U.S., Israel Developed Flame Computer Virus," supra.

90. Ibid.

91. "Flame Virus 'Created by US and Israel as Part of Intensifying Cyber Warfare,'" *The Telegraph*, June 6, 2014.

92. Ibid.

93. Boldizsár Bencsáth, Gábor Pék, Levente Buttyán, and Márk Félegyházi, "Duqu: A Stuxnet-like Malware Found in the Wild," *Laboratory of Cryptography and System Security, Budapest University of Technology and Economics' Department of Telecommunications*, October 2011, http://www.crysys.hu/publications/files/bencsathPBF11duqu.pdf.

94. Ibid.

95. Mark Clayton, "More Telltale Signs of Cyber Spying and Cyberattacks," supra.

96. Ibid.

97. Ibid.
98. Ibid.
99. Ibid.
100. Samantha Bradshaw and Philip N. Howard, "Troops, Trolls and Troublemakers: A Global Inventory of Organized Social Media Manipulation," in Samuel Woolley and Philip N. Howard (eds.), Working Paper, December 2017, (Oxford, UK) Project on Computational Propaganda, http://comprop.oii.ox.ac.uk/research/troops-trolls-and-trouble-makers-a-global-inventory-of-organized-social-media-manipulation/.
101. Craig Timberg, "Spreading Fake News Becomes Standard Practice for Governments across the World," *The Washington Post*, July 17, 2017, https://www.washingtonpost.com/news/the-switch/wp/2017/07/17/spreading-fake-news-becomes-standard-practice-for-governments-across-the-world/?hpid=hp_hp-cards_hp-card-technology%3Ahomepage%2Fcardutm_term=.248d94c26b31.
102. UK House of Commons Digital, Culture, Media and Sport Committee, *Disinformation and 'Fake News': Final Report Eighth Report of Session 2017–19 Report* (London: The House of Commons, 2019), 77.
103. Carole Cadwalladr, "Fresh Cambridge Analytica Leak 'Shows Global Manipulation is Out of Control,'" *The Guardian*, January 4, 2020, https://www.theguardian.com/uk-news/2020/jan/04/cambridge-analytica-data-leak-global-election-manipulation.
104. Patrick Wintour, "Russian Bid to Influence Brexit Vote Detailed in New US Senate Report," *The Guardian*, January 10, 2018, https://www.theguardian.com/world/2018/jan/10/russian-influence-brexit-vote-detailed-us-senate-report.

CHAPTER 2

1. Barton Gellman and Ellen Nakashima, "U.S. Spy Agencies Mounted 231 Offensive Cyber-Operations in 2011, Documents Show," *The Washington Post*, August 30, 2013, https://www.washingtonpost.com/world/national-security/us-spy-agencies-mounted-231-offensive-cyber-operations-in-2011-documents-show/2013/08/30/d090a6ae-119e-11e3-b4cb-fd7ce041d814_story.html?utm_term=.e963d1e06ce4.
2. Nicole Perlroth, Jeff Larson, and Scott Shane, "Revealed: The NSA's Secret Campaign to Crack, Undermine Internet Security," *The New York Times*, September 5, 2013, http://www.propublica.org/article/the-nsas-secret-campaign-to-crack-undermine-internet-encryption.
3. For instance, the United States uses cyber means "to deter, deny, or defeat any adversary that seeks to harm U.S. national interests in *peace*, crisis, or war." (Emphasis added.) See US Presidential Policy Directive/PPD 20 *US Cyber Operations Policy* Washington, DC, 2012, https://fas.org/irp/offdocs/ppd/ppd-20.pdf.
4. White House, National Security Strategy 2017, https://www.whitehouse.gov/wp-content/uploads/2017/12/NSS-Final-12-18-2017-0905.pdf.
5. Ibid.

6. Ibid.
7. Ibid.
8. Ellen Nakashima, "Military Leaders Seek More Clout for Pentagon's Cyber Command Unit," *The Washington Post*, May 1, 2012, https://www.washingtonpost.com/world/national-security/military-officials-push-to-elevate-cyber-unit-to-full-combatant-command-status/2012/05/01/gIQAUud1uT_story.html?utm_term=.82a37e29370b.
9. Office of the Director of National Intelligence, Cyber Threat Intelligence Integration Center, *CTIIC's Mission Responsibilities are Outlined in a February 2015 MEMO PRESIDENTIAL Memorandum that Directed the DNI to Establish CTIIC*, Retrieved July 24, 2017, https://www.dni.gov/files/CTIIC/documents/CTIIC-Overview_for-unclass.pdf.
10. "National Cyber Security Division," *US Department of Homeland Security*. <www.dhs.gov/xabout/structure/editorial_0839.shtm>.
11. See "Homeland Security Presidential Directive 7: Critical Infrastructure Identification, Prioritization, and Protection," December 17, 2003.
12. "National Cyber Security Division," supra.
13. "Cyberspace Policy Review: Assuring a Trusted and Resilient Information and Communications Infrastructure," *The White House*, May 2009, 37.
14. "U.S. Cyber Command Fact Sheet," US Department of Defense. <www.defense.gov/home/features/2010/0410_cybersec/docs/CYberFactSheet%20UPD>.
15. "Cyber Mission Force Achieves Full Operational Capability," U.S. Cyber Command News Release May 17, 2018, https://www.defense.gov/News/Article/Article/1524747/cyber-mission-force-achieves-full-operational-capability/
16. Thom Shanker, "Pentagon Is Updating Conflict Rules in Cyberspace," *The New York Times* June 27, 2013, http://www.nytimes.com/2013/06/28/us/pentagon-is-updating-conflict-rules-in-cyberspace.html.
17. Scott Shane, Mark Mazzetti and Matthew Rosenberg, "WikiLeaks Releases Trove of Alleged C.I.A. Hacking Documents," *New York Times*, March 7, 2017, https://www.nytimes.com/2017/03/07/world/europe/wikileaks-cia-hacking.html?_r=0
18. Jacob Davidson, "China Accuses U.S. of Hypocrisy on Cyberattacks," *Time*, July 1, 2013 http://world.time.com/2013/07/01/china-accuses-u-s-of-hypocrisy-on-cyberattacks/
19. Nakashima, Miller and Tate, "U.S., Israel Developed Flame Computer Virus," supra.
20. Ibid.
21. Gellman and Nakashima, "U.S. Spy Agencies Mounted 231 Offensive Cyber-Operations," supra.
22. Clarke and Robert, *Cyber War*, supra., 179.
23. Nakashima, Miller and Tate, "U.S., Israel Developed Flame Computer Virus," supra.
24. Gil Baram, "The Theft and Reuse of Advanced Offensive Cyber Weapons Pose a Growing Threat," *Council on Foreign Relations*, June 19, 2018, https://www.cfr.org/blog/theft-and-reuse-advanced-offensive-cyber-weapons-pose-growing-threat.

25. White House, *Vulnerabilities Equities Policy and Process for the United States Government* November 15, 2017, https://www.whitehouse.gov/sites/whitehouse.gov/files/images/External%20-%20Unclassified%20VEP%20Charter%20FINAL.PDF

26. Ibid.

27. Franklin D. Kramer, Stuart H. Starr, and Larry Wentz (eds.), *Cyberpower and National Security* (Washington, DC: National Defense University Potomac Books Inc., 2009).

28. "Clarke: More Defense Needed in Cyberspace," *HometownAnnapolis.com*, September 24, 2010.

29. Masters, "Confronting the Cyber Threat," supra.

30. Office of the US President, *The Comprehensive National Cybersecurity Initiative,* 2009.

31. US Cyber Command, *Achieve and Maintain Cyberspace Superiority: Command Vision for US Cyber Command*, March 23, 2018, https://assets.documentcloud.org/documents/4419681/Command-Vision-for-USCYBERCOM-23-Mar-18.pdf.

32. Richard J. Harknett, "United States Cyber Command's New Vision: What It Entails and Why It Matters," *Lawfare*, March 23, 2018, https://www.lawfareblog.com/united-states-cyber-commands-new-vision-what-it-entails-and-why-it-matters.

33. Ibid.

34. Ibid.

35. Ibid.

36. William Clinton, *National Security Strategy* (Washington, DC: Government Printing Office, 1995), 8.

37. White House Presidential Directive/NSC-63, "Critical Infrastructure Protection," May 1998, https://www.documentcloud.org/documents/1513862-clinton-presidential-policy-directive.html

38. This is all with the understanding that DARPA was an early creator of the internet. Also see Keith B. Alexander, "Warfighting in Cyberspace," *Joint Forces Quarterly*, July 31, 2007, http://www.military.com/forums/0,15240,143898,00.html.

39. US White House, *Prosperity, Security, and Openness in a Networked World,* May 11, 2012.

40. US Presidential Executive Order, *Strengthening the Cybersecurity of Federal Networks and Critical Infrastructure* (2017). https://www.whitehouse.gov/the-press-office/2017/05/11/presidential-executive-order-strengthening-cybersecurity-federal

41. "Five strategic initiatives—1. Treat cyberspace as an operational domain to organize, train, and equip so that DoD can take full advantage of cyberspace's potential; 2. Employ new defense operating concepts to protect DoD networks and systems; 3. Partner with other U.S. government departments and agencies and the private sector to enable a whole-of-government cybersecurity strategy; 4. Build robust relationships with U.S. allies and international partners to strengthen collective cybersecurity; and 5. Leverage the nation's ingenuity through an exceptional cyber workforce and rapid technological innovation." *The Department of Defense Strategy for Operating in Cyberspace* (July 2011).

42. US White House, *National Security Strategy*, May 2010.

43. "International Strategy for Cyberspace: Prosperity, Security, and Openness in a Networked World," *The White House*, May 2011, 14 and Lewis and Timlin, *Cybersecurity and Cyberwarfare*, supra.

44. "Department of Defense," *The DOD Cyber Strategy*, April 2015, 13–14. https://www.defense.gov/Portals/1/features/2015/0415_ cyber-strategy/Final_2015_DoD_CYBER_STRATEGY_for_web.pdf.

45. Dustin Volz, "Trump Administration Hasn't Briefed Congress on New Rules for Cyberattacks, Lawmakers Say: Some Lawmakers Are Concerned They Lack Oversight of the Military's Increasing Use of Cyber Weapons," *The Wall Street Journal*, July 10, 2019, https://www.wsj.com/articles/trump-administration-hasnt-briefed-congress-on-new-rules-for-cyberattacks-lawmakers-say-11562787360.

46. Gellman and Nakashima, "U.S. Spy Agencies Mounted 231 Offensive Cyber-Operations," supra.

47. Ibid.

48. Department of Defense/Defense Science Board, Task Force on Cyber Deterrence, February 2017.

49. *The Department of Defense Strategy for Operating in Cyberspace* (July 2011), 6.

50. Department of Defense/Defense Science Board, supra.

51. As quoted in Gellman and Nakashima, "U.S. Spy Agencies Mounted 231 Offensive Cyber Operations," supra.

52. Tom Gjelten, "First Strike: US Cyber Warriors Seize the Offensive," *World Affairs* January/February 2013, http://www.worldaffairsjournal.org/article/first-strike-us-cyber-warriors-seize-offensive

53. DARPA funding enabled the invention of the internet, stealth aircraft, GPS, and voice-recognition software.

54. Gjelten, "First Strike: US Cyber Warriors Seize the Offensive," supra.

55. Gellman and Nakashima, "U.S. Spy Agencies Mounted 231 Offensive Cyber-Operations," supra.

56. "Deterring Hybrid Warfare: "*NATO Review*. http://www.nato.int/docu/review/2014/Also-in-2014/Deterring-hybrid-warfare/EN/index.htm and Lieutenant General James N. Mattis, USMC, and Lieutenant Colonel Frank Hoffman, USMCR (Ret.), "Future Warfare: The Rise of Hybrid Wars," *Proceedings Magazine*, Vol. 132 November 2005, http://milnewstbay.pbworks.com/f/MattisFourBlockWarUSNINov2005.pdf.

57. Coll, "The Rewards (and Risks) of Cyber War," supra.

58. Gjelten, "First Strike: US Cyber Warriors Seize the Offensive," supra.

59. Chinese military strategist Sun Tzu stated, "All warfare is based on deception." See Sun Tzu, The Art of War, ed. and trans. Samuel Griffith (London: Oxford University Press, 1963), 66.

60. Prentiss O. Baker, LTC, USA, "Psychological Operations within the Cyber Domain," *Maxwell Paper No. 52, The Air War College*, February 17, 2010, http://www.au.af.mil/au/awc/awcgate/maxwell/mp52.pdf

61. George F. Kennan, *Policy Planning Staff Memorandum 269*, Washington, DC: U.S. State Department, May 4, 1948. As of May 20, 2019, http://academic.brooklyn.cuny.edu/history/johnson/65ciafounding3.htm.

62. Baker, LTC, USA, "Psychological Operations within the Cyber Domain," supra.

63. Craig Whitlock, "Somali American Caught Up in a Shadowy Pentagon Counterpropaganda Campaign," *The Washington Post*, July 7, 2013, https://www

.washingtonpost.com/world/national-security/somali-american-caught-up-in-a-shadowy-pentagon-counterpropaganda-campaign/2013/07/07/b3aca190-d2c5-11e2-bc43-c404c3269c73_story.html.

64. Shanker, "Pentagon Is Updating Conflict," supra.

65. Ibid.

66. Glenn Greenwald and Ewen MacAskill, "Obama Orders US to Draw Up Overseas Target List for Cyber-Attack," *The Guardian*, June 7, 2013, http://www.theguardian.com/world/2013/jun/07/obama-china-targets-cyber-overseas

67. Department of Defense/Defense Science Board, supra.

68. Greg Miller, Ellen Nakashima and Adam Entous, "Obama's Secret Struggle to Punish Russia for Putin's Election Assault," *The Washington Post*, June 23, 2017, https://www.washingtonpost.com/graphics/2017/world/national-security/obama-putin-election-hacking/?tid=ss_mailutm_term=.383942a7e764.

69. Department of Defense/Defense Science Board, supra.

70. Ibid.

71. Staff Study Prepared in the Department of Defense, "Evaluation of Possible Military Courses of Action," *Foreign Relations of the United States, 1961-63, Volume X Cuba, January 1961–September 1962,* https://history.state.gov/historicaldocuments/frus1961-63v10/d19.

72. Central Intelligence Agency, Studies in Intelligence: A collection of articles on the historical, operational, doctrinal, and theoretical aspects of intelligence Winter 1999-2000 Unclassified Edition https://cryptome.org/nsa-shamrock.htm.

73. Barton Gellman, Ashkan Soltani, and Andrea Peterson, "How We Know the NSA had Access to Internal Google and Yahoo Cloud Data," *The Washington Post* November 4, 2013, http://www.washingtonpost.com/blogs/the-switch/wp/2013/11/04/how-we-know-the-nsa-had-access-to-internal-google-and-yahoo-cloud-data/?tid=hpModule_88854bf0-8691-11e2-9d71-f0feafdd1394hpid=z11

74. Ryan Lizza, "State of Deception--Why won't the President Rein in the Intelligence Community?" *New Yorker*, December 16, 2013, http://www.newyorker.com/reporting/2013/12/16/131216fa_fact_lizza?currentPage=all and Gellman, Soltani, and Peterson, "How We Know the NSA had Access," supra.

75. Offices of Inspectors General of the Department of Defense, Department of Justice, Central Intelligence Agency, National Security Agency, and Office of the Director of National Intelligence, Annex to the Report on the President's Surveillance Program Volume III July 10, 2009, https://oig.justice.gov/reports/2015/PSP-09-18-15-vol-III.pdf.

76. Ellen Nakashima, "Legal Memos Released on Bush-Era Justification For Warrantless Wiretapping," *The Washington Post* September 6, 2014, https://www.washingtonpost.com/world/national-security/legal-memos-released-on-bush-era-justification-for-warrantless-wiretapping/2014/09/05/91b86c52-356d-11e4-9e92-0899b306bbea_story.html.

77. Ibid.

78. Dustin Volz, "Trump Signs Bill Renewing NSA's Internet Surveillance Program," *Reuters* January 19, 2018, https://www.reuters.com/article/us-usa-trump-cyber-surveillance/trump-signs-bill-renewing-nsas-internet-surveillance-program-idUSKBN1F82MK

79. Department of Defense/Defense Science Board, supra.

80. Gellman and Ellen, "U.S. Spy Agencies Mounted 231 Offensive Cyber-Operations," supra.

81. David E. Sanger, and Nicole Perlroth, "N.S.A. Breached Chinese Servers Seen as Security Threat," *The New York Times*, March 22, 2014, https://www.nytimes.com/2014/03/23/world/asia/nsa-breached-chinese-servers-seen-as-spy-peril.html.

82. Jack Goldsmith, "The Precise (and Narrow) Limits on U.S. Economic Espionage," *Lawfare* March 23, 2015, https://www.lawfareblog.com/precise-and-narrow-limits-us-economic-espionage.

83. Ibid.

84. Ibid.

85. Ibid.

86. Davidson, "China Accuses U.S. of Hypocrisy," supra

87. Ibid.

88. Gellman and Nakashima, "U.S. Spy Agencies Mounted 231 Offensive Cyber-Operations," supra.

89. Greenwald and MacAskill, "Obama Orders US to Draw Up Overseas Target List for Cyber-Attacks," supra.

90. Gellman and Nakashima, "U.S. Spy Agencies Mounted 231 Offensive Cyber-Operations," supra.

91. Department of Defense/Defense Science Board, supra.

92. Shane, Mazzetti and Rosenberg, "WikiLeaks Releases Trove," supra.

93. Gellman and Nakashima, "U.S. Spy Agencies Mounted 231 Offensive Cyber-Operations," supra.

94. As quoted in Perlroth, Larson, and Shane, "Revealed: The NSA's Secret Campaign," supra.

95. Perlroth, Larson, and Shane, "Revealed: The NSA's Secret Campaign," supra.

96. Jenny Jun, Scott LaFoy, and Ethan Sohn, "North Korea's Cyber Operations: Strategy and Responses," December 2015, The Center for Strategic and International Studies (CSIS), 22.

97. "Big Surge in Cyberattacks on Russia amid US Hacking Hysteria—Russian Security Chief," *Russia Today*, January 15, 2017, https://www.rt.com/news/373764-surge-hacking-attacks-russia/.

98. David J. Smith, "Russian Cyber Capabilities, Policy and Practice," in *Focus Quarterly* Winter 2014.

99. In addition to this quote, this article asserts that "Vladimir Putin, who is quick to accuse the West of hypocrisy, frequently points to this history. He sees a straight line from the West's support of the anti-Moscow 'color revolutions,' in Georgia, Kyrgyzstan, and Ukraine, which deposed corrupt, Soviet-era leaders, to its endorsement of the uprisings of the Arab Spring." See Evan Osnos, David Remnick, and Joshua Yaffa, "Trump, Putin and the New Cold War, "*New Yorker*, March 6, 2017, http://www.newyorker.com/magazine/2017/03/06/trump-putin-and-the-new-cold-war.

100. Mischa Gabowitsch, *Protest in Putin's Russia* (Cambridge: Polity Press, 2017), Chap. 1.

101. Baker, LTC, USA, "Psychological Operations within the Cyber Domain," supra.

102. Kennon H. Nakamura and Matthew C Weed, US Public Diplomacy: Background and Current Issues (Washington DC: Congressional Research Service December 18, 2009), 16.

103. Ibid.

104. Ibid., 37.

105. Ibid., 52.

106. Kartanarusheniy October 8, 2019, https://www.kartanarusheniy.org.

107. "Vladimir Putin, who is quick to accuse the West of hypocrisy, frequently points to this history. He sees a straight line from the West's support of the anti-Moscow 'color revolutions,' in Georgia, Kyrgyzstan, and Ukraine, which deposed corrupt, Soviet-era leaders, to its endorsement of the uprisings of the Arab Spring." Osnos, Remnick and Yaffa, "Trump, Putin and the New Cold War," supra.

108. Norwegian Helsinki Committee https://www.nhc.no/en/frontpage/

109. As quoted in "Election watchdog Golos demands to be removed from 'foreign agents' list after court victory," *Russia Today*, September 9, 2014, https://www.rt.com/politics/186452-golos-watchdog-ngo-court/.

110. David E. Sanger and Nicole Perlroth, "U.S. Escalates Online Attacks on Russia's Power Grid," *New York Times*, June 15, 2019, https://www.nytimes.com/2019/06/15/us/politics/trump-cyber-russia-grid.html.

111. Ken Bredemeier, "Russia Demands Return of 2 Shuttered Compounds in US," *Voice of America*, July 17, 2017, https://www.voanews.com/usa/russia-demands-return-2-shuttered-compounds-us.

112. Options discussed for deterring or punishing Russia included cyberattacks on Russian infrastructure, the release of CIA-gathered material to embarrass Putin, and sanctions to damage the Russian economy, according to Greg Miller, Ellen Nakashima, and Adam Entous, "Obama's Secret Struggle to Punish Russia," supra.

113. Ibid.

114. The five warnings were: (1) an August 4, 2016, phone call from CIA director John Brennan to the director of the FSB (the successor to the KGB) Alexander Bortnikov; (2) a September 2016 confrontation between Obama and Putin during a meeting of world leaders in Hangzhou, China; (3) a September 5, 2016, news conference, where Obama issued a veiled threat; (4) an October 7, 2016, message to Putin relayed by US National Security Adviser Susan Rice to Russian ambassador Sergey Kislyak; and (5) on October 31, 2016, the administration delivered a final preelection message via a secure nuclear war-era channel. See Miller, Nakashima and Entous, "Obama's Secret Struggle to Punish Russia," supra.

115. Ibid.

116. Ibid.

117. Robert Tait and Julian Borger, "Alleged Hacker Held in Prague at Center of 'Intense' US-Russia Tug of War," *The Guardian*, January 27, 2017, https://www.theguardian.com/technology/2017/jan/27/us-russia-hacking-yevgeniy-nikulin-linkedin-dropbox

118. Kartikay Mehrotra, "Jailed Russian of Interest in U.S. Election Probe, Official Says," *Bloomberg News*, August 24, 2018, https://www.bloomberg.com/news/articles/2018-08-24/quiet-jailed-russian-said-of-interest-in-u-s-election-meddling.

119. Tait and Borger, "Alleged Hacker Held in Prague," supra.

120. "Russian Computer Programmer Arrested in Spain: Embassy," *Reuters*, April 9, 2017, http://www.reuters.com/article/us-spain-russia-idUSKBN17B0O2.

121. Reuters Staff, "Russian Accused of Hacking Extradited to U.S. from Spain," *Reuters* February 2, 2018, https://www.reuters.com/article/us-usa-cyber-levashov/russian-accused-of-hacking-extradited-to-u-s-from-spain-idUSKBN1FM2RG.

122. US Indictment CRIMINAL NO. (18 U.S.C. §§ 2, 371, 1349, 1028A) February 16, 2018, https://www.justice.gov/file/1035477/download.

123. Mark Mazzetti and Katie Benner, "12 Russian Agents Indicted in Mueller Investigation," *The New York Times*, July 13, 2018, https://www.nytimes.com/2018/07/13/us/politics/mueller-indictment-russian-intelligence-hacking.html.

124. Julian E. Barnes and Thomas Gibbons-Neff, "U.S. Carried Out Cyberattacks on Iran," *New York Times*, June 22, 2019, https://www.nytimes.com/2019/06/22/us/politics/us-iran-cyber-attacks.html.

125. Michael Shear, Eric Schmitt, and Maggie Haberman, "Trump Approves Strike on Iran, but then Abruptly Pulls Back," *The New York Times*, June 20, 2019, https://www.nytimes.com/2019/06/20/world/middleeast/iran-us-drone.html.

126. David E. Sanger and Eric Schmitt, "U.S. Cyberweapons, Used Against Iran and North Korea, Are a Disappointment Against ISIS," *New York Times* June 12, 2017, https://www.nytimes.com/2017/06/12/world/middleeast/isis-cyber.html?smid=nytcore-ipad-sharesmprod=nytcore-ipad_r=0

127. Sanger and Schmitt, "U.S. Cyberweapons, Used against Iran," supra.

128. David E. Sanger and William J Broad, "Hand of U.S. Leaves North Korea's Missile Program Shaken," *New York Times*, April 18, 2017, https://www.nytimes.com/2017/04/18/world/asia/north-korea-missile-program-sabotage.html?_r=0

129. Karen DeYoung, Ellen Nakashima and Emily Rauhala, "US-Trump signed presidential directive ordering actions to pressure North Korea," *The Washington Post* September 30, 2017, https://www.washingtonpost.com/world/national-security/trump-signed-presidential-directive-ordering-actions-to-pressure-north-korea/2017/09/30/97c6722a-a620-11e7-b14f-f41773cd5a14_story.html?utm_term=.5d67f8804aa6

130. Sanger and Schmitt, "U.S. Cyberweapons, Used Against Iran," supra.

131. Ibid.

132. Ibid.

133. Gellman and Nakashima, "U.S. Spy Agencies Mounted 231 Offensive Cyber-Operations," supra.

134. Siobhan Gorman, "Electricity Grid in U.S. Penetrated By Spies," *The Wall Street Journal*, April 8, 2009, http://online.wsj.com/article/SB123914805204099085.html.

135. Ibid.

136. "The security intelligence firm CrowdStrike associated the Havex RAT with targeted attacks against energy sector organizations in September 2013, which were

perpetrated by a group of attackers with links to Russia. The security firm dubbed the attack group 'Energetic Bear' and said that its malicious campaigns go as far back as August 2012." Lucian Constantin, "New Havex malware variants target industrial control system and SCADA users," *PC World* June 24, 2014 http://www.pcworld.com/article/2367240/new-havex-malware-variants-target-industrial-control-system-and-scada-users.html and Havex Trojan: ICS-ALERT-14-176-02A

137. This malware from Russia was used to shut down the parts of the Ukraine power grid, the first known attack to do so, in December 23, 2014. BlackEnergy: ICS-ALERT-14-281-01E

138. Department of Defense/Defense Science Board, supra.

139. US White House, *Prosperity, Security, and Openness*, supra.

140. Ellen Nakashima, "With Plan X, Pentagon Seeks to Spread U.S. Military Might to Cyberspace," *The Washington Post*, May 30, 2012, http://www.washingtonpost.com/world/national-security/with-plan-x-pentagon-seeks-to-spread-us-military-might-to-cyberspace/2012/05/30/gJQAEca71U_story.html

141. Ibid.

142. US Office of the director of US Intelligence, *Background to "Assessing Russian Activities and Intentions in Recent US Elections": The Analytic Process and Cyber Incident Attribution* January 6, 2017.

143. Larry Downes, "On Internet Regulation, The FCC Goes Back To The Future," *Forbes* March 12, 2018, https://www.forbes.com/sites/larrydownes/2018/03/12/the-fcc-goes-back-to-the-future/#56d1e3d05b2e

144. Keith Collins, "Net Neutrality Has Officially Been Repealed. Here's How That Could Affect You," *The New York Times*, June 11, 2018, https://www.nytimes.com/2018/06/11/technology/net-neutrality-repeal.html.

145. Downes, "On Internet Regulation," supra.

146. Collins, "Net Neutrality Has Officially Been Repealed." supra.

147. US Joint Force Development, "Cyberspace Operations," June 8, 2018, https://www.jcs.mil/Portals/36/Documents/Doctrine/pubs/jp3_12.pdf?ver=2018-07-16-134954-150.

148. Ibid.

149. Eneken Tikk and Mika Kerttunen, "Parabasis: Cyber-diplomacy in Stalemate," *Norwegian Institute of International Affairs*, May 2018, https://nupi.brage.unit.no/nupi-xmlui/bitstream/handle/11250/2569401/NUPI_Report_5_18_Tikk_Kerttunen.pdf?sequence=1isAllowed=y

150. *Developments in the Field of Information and Telecommunications in the Context of International Security* (A/54/213).

151. Tikk and Kerttunen, "Parabasis: Cyber-diplomacy in Stalemate," supra.

152. Julian Ku, "Forcing China to Accept that International Law Restricts Cyber Warfare May Not Actually Benefit the U.S.," Lawfare (August 25, 2017).

153. David E. Sanger and William J Broad, "Pentagon Suggests Countering Devastating Cyberattacks with Nuclear Arms," *The New York Times*, January 16, 2018, https://www.nytimes.com/2018/01/16/us/politics/pentagon-nuclear-review-cyberattack-trump.html.

154. Ku, "Forcing China to Accept that International Law Restricts Cyber Warfare," supra.

155. Sanger and Broad, "Pentagon Suggests Countering Devastating," supra.

156. John Markoff and Andrew E. Kramer, "U.S. and Russia Differ on a Treaty for Cyberspace," *New York Times*, June 27, 2009, http://www.nytimes.com/2009/06/28/world/28cyber.html.

157. US White House, *Prosperity, Security, and Openness*, supra.

158. Ibid.

159. Carl von Clausewitz, *On War* Book 1, Chapter 1, 1873, https://www.clausewitz.com/readings/OnWar1873/BK1ch01.html.

160. René Girard, *On War and Apocalypse*, August 2009, https://www.firstthings.com/article/2009/08/on-war-and-apocalypse.

161. Randall R. Dipert, The Ethics of Cyberwarfare, *Journal of Military Ethics*, 9:4, 384–410. 2010, http://dx.doi.org/10.1080/15027570.2010.536404.

162. Col. James Cook, "'Cyberation' and Just War Doctrine: A Response to Randall Dipert," *Journal of Military Ethics*, 9:4, 411–423, 2010, https://www.law.upenn.edu/live/files/1701-cook-cyberation

163. US White House, *Prosperity, Security, and Openness*, supra.

CHAPTER 3

1. Gellman and Nakashima, "U.S. Spy Agencies Mounted 231 Offensive Cyber-Operations," supra.

2. Jānis Bērziņš, "Russia's New Generation Warfare in Ukraine," *National Defence Academy of Latvia Center for Security and Strategic Research Policy Paper No. 2*, April 2014, http://www.sldinfo.com/wp-content/uploads/2014/05/New-Generation-Warfare.pdf.

3. "Военная доктрина Российской Федерации," Russian Presidential Executive Office, February 5, 2010, http://news.kremlin.ru/ref_notes/461.

4. "Cyber Wars," *Agentura.Ru*, <www.agentura.ru/english/equipment/>.

5. Decree of the President of the Russian Federation, "On approval of the Doctrine of Information Security of the Russian Federation," December 5, 2016, http://publication.pravo.gov.ru/Document/View/0001201612060002.

6. Sergey Sukhankin, "Russia's New Information Security Doctrine: Fencing Russia from the 'Outside World'?" *Eurasia Daily Monitor* 13:198, 2016, https://www.refworld.org/docid/5864c6b24.html.

7. Constanze Stelzenmüller's testimony before the US Senate Select Committee on Intelligence, "The Impact of Russian Interference on Germany's 2017 Election," *Brookings Institute*, June 28, 2017, https://www.brookings.edu/testimonies/the-impact-of-russian-interference-on-germanys-2017-elections/

8. As quoted in Von Patrick Beuth, Kai Biermann, Martin Klingst und Holger Stark, "Cyberattack on the Bundestag: Merkel and the Fancy Bear," *Zeit Online* May 12, 2017, https://www.zeit.de/digital/2017-05/cyberattack-bundestag-angela-merkel-fancy-bear-hacker-russia.

9. Vladimir Putin, *Russia and the Changing World*, February 27, 2012, https://www.rt.com/politics/official-word/putin-russia-changing-world-263/.

10. H. A. Conley, J. Mina, R. Stefanov, M. Vladimirov, *The Kremlin Playbook: Understanding Russian Influence in Central and Eastern Europe* (Boulder CO: Rowman Littlefield, 2016), iv–v.

11. Mark Galeotti, "Crimintern: How the Kremlin Uses Russia's Criminal Networks in Europe," *European Council on Foreign Relations*, April 18, 2017, https://www.ecfr.eu/publications/summary/crimintern_how_the_kremlin_uses_russias_criminal_networks_in_europe.

12. Mark Galeotti, "The Kremlin's Newest Hybrid Warfare Asset," *Foreign Policy*, June 12, 2017, https://foreignpolicy.com/2017/06/12/how-the-world-of-spies-became-a-gangsters-paradise-russia-cyberattack-hack/.

13. Galeotti, "Crimintern: How the Kremlin Uses Russia's," supra.

14. Stelzenmüller, "The Impact of Russian Interference on Germany's 2017 Election," supra.

15. "WannaCry: Are You Safe?" *Kaspersky Company*, May 15, 2017, https://www.kaspersky.co.uk/blog/wannacry-ransomware/8700/.

16. Eugene Gerden, "$500 Million for New Russian Cyber Army," *SC Magazine*, November 6, 2014, http://www.scmagazineuk.com/500-million-for-new-russian-cyberarmy/article/381720/.

17. Smith, "Russian Cyber Capabilities," supra.

18. Lewis and Timlin, *Cybersecurity and Cyberwarfare*, supra.

19. Gerden, "$500 Million for New Russian Cyber Army," supra.

20. Roland Heickerö, "Emerging Cyber Threats and Russian Views on Information Warfare and Information Operations," *Swedish Defence Research Agency*, 2010, 27ff.

21. Andrea Shalal, "Germany Challenges Russia over Alleged Cyberattacks," *Reuters*, May 4, 2017, http://www.reuters.com/article/us-germany-security-cyber-russia-idUSKBN1801CA.

22. Kevin Kelleher, "Microsoft Says Russia Has Already Tried to Hack 3 Campaigns in the 2018 Election," *Fortune*, July 19, 2018, http://fortune.com/2018/07/19/microsoft-russia-hack-2018-election-campaigns/.

23. Garrett M. Graff, "Indicting 12 Russian Hackers Could Be Mueller's Biggest Move Yet," *Wired*, July 13, 2018, https://www.wired.com/story/mueller-indictment-dnc-hack-russia-fancy-bear/.

24. "Dutch Intelligence First to Alert U.S. about Russian Hack of Democratic Party," *Nieuwsuur*, January 25, 2018, https://nos.nl/nieuwsuur/artikel/2213767-dutch-intelligence-first-to-alert-u-s-about-russian-hack-of-democratic-party.html.

25. Mikhail Barabanov, "Testing a "New Look,'" *Russia in Global Affairs*, December 18, 2014, https://eng.globalaffairs.ru/number/Testing-a-New-Look-17213.

26. Andrew Monaghan, "Putin's Way of War: The 'War' in Russia's 'Hybrid Warfare,'" *Parameters*, Winter 2015–2016, 66 and 70. https://ssi.armywarcollege.edu/pubs/parameters/issues/winter_2015-16/9_monaghan.pdf.

27. Gerden, "$500 Million for New Russian Cyber Army," supra.

28. Vasudevan Sridharan. "Russia Setting Up Cyber Warfare Unit Under Military," *International Business Times*, August 20, 2013, http://www.ibtimes.co.uk/articles/500220/20130820/russia-cyber-war-hack-moscow-military-snowden.htm.

29. Gerden, "$500 Million for New Russian Cyber Army," supra.

30. The Grand Jury Indictment states that "twelve of the individual defendants worked at various times for Internet Research Agency LLC, a Russian company based in St. Petersburg, Russia. The other individual defendant, Yevgeniy Viktorovich Prigozhin, funded the conspiracy through companies known as Concord Management and Consulting LLC, Concord Catering, and many subsidiaries and affiliates. . . . Internet Research Agency allegedly operated through Russian shell companies. It employed hundreds of persons for its online operations, ranging from creators of fictitious personas to technical and administrative support, with an annual budget of millions of dollars. Internet Research Agency was a structured organization headed by a management group and arranged in departments, including graphics, search-engine optimization, information technology, and finance departments." Office of Public Affairs, Department of Justice, "Grand Jury Indicts Thirteen Russian Individuals and Three Russian Companies for Scheme to Interfere in the United States Political System," *Justice News*, February 16, 2018, https://www.justicc.gov/opa/pr/grand-jury-indicts-thirteen-russian-individuals-and-three-russian-companies-scheme-interfere.

31. Ibid.

32. Brian Barrett, "For Russia, Unravelling US Democracy Was Just Another Day Job," *Wired*, February 17, 2018, https://www.wired.com/story/mueller-indictment-internet-research-agency/.

33. Ivan Nechepurenko and Michael Schwirtz, "What We Know About Russians Sanctioned by the United States," *The New York Times*, February 17, 2018, https://www.nytimes.com/2018/02/17/world/europe/russians-indicted-mueller.html.

34. "This 'information war,' said Rastislav Kacer, a veteran diplomat who served as Slovakia's ambassador to Washington and at NATO's headquarters in Brussels, 'is just part of a bigger struggle.' While not involving bloodshed, he added, it 'is equally as dangerous as more conventional hostile action.'" As quoted from Andrew Higgins, "Effort to Expose Russia's 'Troll Army' Draws Vicious Retaliation," *New York Times*, May 31, 2016, http://www.nytimes.com/2016/05/31/world/europe/russia-finland-nato-trolls.html?_r=0.

35. Smith, "Russian Cyber Capabilities," supra.

36. Conley, Mina, Stefanov, and Vladimirov, *The Kremlin Playbook*, supra., x.

37. Ibid., xiv.

38. Ibid., x.

39. Ibid.

40. Ibid.

41. Robert Windham, "Timeline: Ten Years of Russian Cyber Attacks on Other Nations," *NBC News*, December 18, 2016, http://www.nbcnews.com/news/us-news/timeline-ten-years-russian-cyber-attacks-other-nations-n697111.

42. Andrew Kramer, "Russian General Pitches 'Information' Operations as a Form of War," *New York Times*, March 2, 2019, https://www.nytimes.com/2019/03/02/world/europe/russia-hybrid-war-gerasimov.html.

43. According to Clapper's testimony, "Computer security studies assert that Russian cyber actors are developing means to remotely access industrial control systems (ICS) used to manage critical infrastructures. Unknown Russian actors successfully compromised the product supply chains of at least three ICS vendors so that customers downloaded malicious software ('malware') designed to facilitate exploitation directly from the vendors' websites along with legitimate software updates, according to private sector cyber security experts." See James R. Clapper, "Statement for the Record: Worldwide Cyber Threats," *House Permanent Select Committee on Intelligence*, September 10, 2015, https://www.dni.gov/index.php/newsroom/testimonies/209-congressional-testimonies-2015/1251-dni-clapper-statement-for-the-record,-worldwide-cyber-threats-before-the-housepermanent-select-committee-on-intelligence.

44. Michael Connell and Sarah Vogler, "Russia's Approach to Cyber Warfare," *CNA Analysis and Solutions*, March 2017, 27–29. https://www.cna.org/CNA_files/PDF/DOP-2016-U-014231-1Rev.pdf.

45. Ibid.

46. Bērziņš, "Russia's New Generation Warfare in Ukraine," supra.

47. Amos Chapple, "The Art of War: Russian Propaganda in WWI," *Radio Free Europe/Radio Liberty*, 2018, https://www.rferl.org/a/russias-world-war-one-propaganda-posters/29292228.html.

48. Phillip Karber and Joshua Thibeault, "Russia's New-Generation Warfare," *Association of the United States Army*, May 20, 2016, https://www.ausa.org/articles/russia%E2%80%99s-new-generation-warfare.

49. Nathalie Maréchal, "Are You Upset About Russia Interfering With Elections?" *Slate*, March 20, 2017, http://www.slate.com/articles/technology/future_tense/2017/03/russia_s_election_interfering_can_t_be_separated_from_its_domestic_surveillance.html.

50. "Report of the International Agora 'Freedom of the Internet 2018: Delegation of Repression,'" May 2, 2019, https://www.agora.legal/articles/Doklad-Mezhdunarodnoi-Agory-%C2%ABSvoboda-interneta-2018-delegirovanie-repressiy%C2%BB/18.

51. Amy Mackinnon, "Tinder and the Russian Intelligence Services: It's a Match! Will Facebook and Twitter be Next?" *Foreign Policy*, June 7, 2019, https://foreignpolicy.com/2019/06/07/tinder-and-the-russian-intelligence-services-its-a-match/.

52. Maréchal, "Are You Upset About Russia Interfering With Elections?" supra.

53. Smith, "Russian Cyber Capabilities," supra.

54. "Russia Enacts 'Draconian' Law for Bloggers and Online Media," *BBC News*, August 1, 2014, https://www.bbc.com/news/technology-28583669.

55. Freedom House, "Freedom on the Net 2015," *Freedom House*, October 2015, 653. https://freedomhouse.org/sites/default/files/FOTN%202015%20Full%20Report.pdf.

56. Adrian Chen, "The Agency," *The New York Times*, June 2, 2015, https://www.nytimes.com/2015/06/07/magazine/the-agency.html.

57. Stelzenmüller, "The Impact of Russian Interference on Germany's 2017 Election," supra.

58. "Russia Internet: Law Introducing New Controls Comes into Force," *BBC News*, November 1, 2019, https://www.bbc.com/news/world-europe-50259597.

59. Cliff Saran, "F-Secure Warns of Russian State-Supported Cyber Espionage," *Computer Weekly*, September 17, 2015, http://www.computerweekly.com/news/4500253704/F-Secure-warns-of-Russian-state-supported-cyber-espionage.

60. Dave Lee, "Red October Cyber-Attack Found by Russian Researchers," *BBC News*, January 14, 2013, http://www.bbc.com/news/technology-21013087.

61. Bob Drogin, "Russians Seem to Be Hacking into Pentagon/Sensitive Information Taken—But Nothing Top Secret," *Los Angeles Times*, October 7, 1999.

62. Scott Shane, Nicole Perlroth, and David E. Sanger, "Security Breach and Spilled Secrets Have Shaken the N.S.A. to Its Core," *The New York Times*, November 12, 2017, https://www.nytimes.com/2017/11/12/us/nsa-shadow-brokers.html.

63. Ibid.

64. Edward Snowden, *Twitter*, August 16, 2016. https://twitter.com/snowden/status/765513776372342784?lang=en.

65. Andrew Morse, "Snowden: Alleged NSA attack is Russian warning," *cnet*, August 16, 2016, https://www.cnet.com/news/snowden-nsa-hack-russia-warning-election-democratic-party/.

66. Snowden, *Twitter*, supra.

67. Windham, "Timeline: Ten Years of Russian Cyber Attacks on Other Nations," supra.

68. Roland Oliphant, Rory Mulholland, Justin Huggler, Senay Boztas, "How Vladimir Putin and Russia are Using Cyberattacks and Fake News to Try to Rig Three Major European Elections this Year," *The Telegraph*, February 13, 2017, http://www.telegraph.co.uk/news/2017/02/13/vladimir-putin-russia-using-cyber-attacks-fake-news-try-rig/.

69. Shaun Waterman, "Analysis: Who Cyber Smacked Estonia?" *UPI*, June 11, 2007, http://www.upi.com/Business_News/Security-Industry/2007/06/11/Analysis-Who-cyber-smacked-Estonia/UPI-26831181580439/.

70. "Estonia Hit by 'Moscow Cyber War,'" *BBC News*, May 17, 2007, http://news.bbc.co.uk/go/pr/fr/-/2/hi/europe/6665145.stm.

71. Ibid.

72. Windham, "Timeline: Ten Years of Russian Cyber Attacks on Other Nations," supra.

73. John Markoff, "Before the Gunfire, Cyberattacks," *The New York Times*, August 12, 2008, http://www.nytimes.com/2008/08/13/technology/13cyber.html?em_r=0.

74. Ibid.

75. Travis Wentworth, "How Russia May Have Attacked Georgia's Internet," *Newsweek*, August 22, 2008, http://www.newsweek.com/how-russia-may-have-attacked-georgias-internet-88111.

76. Markoff, "Before the Gunfire, Cyberattacks," supra.

77. Windham, "Timeline: Ten Years of Russian Cyber Attacks on Other Nations," supra.

78. Ibid.

79. Ibid.

80. Reuters, "Russian Hackers Accused of Targeting UN Chemical Weapons Watchdog, MH17 Files," *Australian Broadcasting Corporation*, October 4, 2018, https://www.abc.net.au/news/2018-10-04/russia-tried-to-hack-un-chemical-weapons-watchdog-netherlands/10339920.

81. Windham, "Timeline: Ten Years of Russian Cyber Attacks on Other Nations," supra.

82. Ibid.

83. Ibid.

84. Greenberg, "How an Entire Nation Became Russia's Test Lab for Cyberwar," supra.

85. Gabe Joselow, "Election Cyberattacks: Pro-Russia Hackers Have Been Accused in Past," *NBC News*, November 3, 2016, https://www.nbcnews.com/mach/technology/election-cyberattacks-pro-russia-hackers-have-been-accused-past-n673246.

86. Ellen Nakashima, "Russia Has Developed a Cyberweapon that Can Disrupt Power Grids, According to New Research," *The Washington Post*, June 12, 2017, https://www.washingtonpost.com/world/national-security/russia-has-developed-a-cyber-weapon-that-can-disrupt-power-grids-according-to-new-research/2017/06/11/b91b773e-4eed-11e7-91eb-9611861a988f_story.html?tid=ss_mailutm_term=.35bafd178f13.

87. Ibid.

88. US Department of Homeland Security, ICS-CERT, "Alert (ICS-ALERT-17-206-01) CRASHOVERRIDE Malware," July 25, 2017, https://ics-cert.us-cert.gov/alerts/ICS-ALERT-17-206-01.

89. Nakashima, "Russia has Developed a Cyberweapon that can Disrupt Power Grids," supra.

90. Greenberg, "How an Entire Nation Became Russia's Test Lab for Cyberwar," supra.

91. Sam Sokol, "Russian Disinformation Distorted Reality in Ukraine. Americans Should Take Note," *Foreign Policy*, August 2, 2019, https://foreignpolicy.com/2019/08/02/russian-disinformation-distorted-reality-in-ukraine-americans-should-take-note-putin-mueller-elections-antisemitism/.

92. US Office of the Director of US Intelligence, *Background to "Assessing Russian Activities and Intentions in Recent US Elections": The Analytic Process and Cyber Incident Attribution* January 6, 2017.

93. Greenberg, "How an Entire Nation Became Russia's Test Lab for Cyberwar," supra.

94. US Office of the Director of US Intelligence, *Background to "Assessing Russian Activities and Intentions in Recent US Elections*," supra.

95. US Senate, Committee on Intelligence, "Report of the Select Committee on Intelligence of the United States Senate on Russian Active Measures Campaigns and Interference in the 2016 US Election," Volume 4, Review of the Intelligence Community Assessment with Additional Views Washington, DC, April 2020, https://www.intelligence.senate.gov/sites/default/files/documents/Report_Volume4.pdf.

96. Warwick Ashford, "Security Research Links Russia to US Election Cyberattacks: Security Researchers Say the Hacking of the US Democratic National Convention's Email System is Linked to a Wider Russian Cyber Campaign," *Computer Weekly*, January 6, 2017, http://www.computerweekly.com/news/450410516/Security-research-links-Russia-to-US-election-cyber-attacks.

97. Michael Riley and Jordan Robertson, "Russian Cyber Hacks on U.S. Electoral System Far Wider Than Previously Known," *Bloomberg News*, June 13, 2017, https://www.bloomberg.com/news/articles/2017-06-13/russian-breach-of-39-states-threatens-future-u-s-elections.

98. Robert S. Mueller, III, Rosalind S Helderman; Matt Zapotosky; US Department of Justice. Special Counsel's Office, *Report on the Investigation into Russian Interference in the 2016 Presidential Election* (New York: Scribner, 2019).

99. Frances Robles, "Russian Hackers Were 'In a Position' to Alter Florida Voter Rolls, Rubio Confirms," *New York Times*, April 26, 2019, https://www.nytimes.com/2019/04/26/us/florida-russia-hacking-election.html.

100. US Office of the Director of US Intelligence, *Background to "Assessing Russian Activities and Intentions in Recent US Elections,"* supra.

101. Steve Dent, "Report: Russia Hacked Election Systems in 39 US States," *Engadget*, June 13, 2017, https://www.engadget.com/2017/06/13/report-russia-hacked-election-systems-in-39-us-states/.

102. Del Quentin Wilber, "Contractor Accused of Leaking NSA Document on Russian Hacking Pleads Guilty," *The Wall Street Journal*, June 26, 2018, https://www.wsj.com/articles/contractor-accused-of-leaking-nsa-document-on-russian-hacking-pleads-guilty-1530048276.

103. Matthew Cole, Richard Esposito, Sam Biddle, and Ryan Grim, "Top-Secret NSA Report Details Russian Hacking Effort Days Before 2016 Election," *The Intercept*, June 5, 2017, https://theintercept.com/2017/06/05/top-secret-nsa-report-details-russian-hacking-effort-days-before-2016-election/.

104. Ibid.

105. Ibid.

106. Ibid.

107. "Hackers Gain Entry into US, European Energy Sector, Symantec Warns," *Reuters*, September 6, 2017, https://www.cnbc.com/2017/09/06/hackers-gain-entry-into-us-european-energy-sector-symantec-warns.html.

108. Joel Schectman, Dustin Volz, and Jack Stubbs, "HP Enterprise Let Russia Scrutinize Cyberdefense System Used by Pentagon," *Reuters*, October 2, 2017, http://www.reuters.com/article/us-usa-cyber-russia-hpe-specialreport/special-report-hp-enterprise-let-russia-scrutinize-cyberdefense-system-used-by-pentagon-idUSKCN1C716M.

109. Robert Tait, "Czech Cyber-Attack: Russia Suspected of Hacking Diplomats' Emails," *The Guardian*, January 31, 2017, https://www.theguardian.com/world/2017/jan/31/czech-cyber-attack-russia-suspected-of-hacking-diplomats-emails.

110. James Shotter, "Czechs Fear Russian Fake News in Presidential Election," *Financial Times* January 8, 2018, https://www.ft.com/content/c2b36cf0-e715-11e7-8b99-0191e45377ec.

111. Tait, "Czech Cyber-Attack," supra.

112. Matthew Czekaj, "Russia's Hybrid War Against Poland," *Eurasia Daily Monitor* 12:80 April 29, 2015, https://jamestown.org/program/russias-hybrid-war-against-poland/.

113. Sabine Fischer, "The Donbas Conflict: Opposing Interests and Narratives, Difficult Peace Process," *SWP Research Paper* 2019/RP 05, April 2019, https://www.swp-berlin.org/10.18449/2019RP05/.

114. NATO, "Statement of the NATO-Ukraine Commission," North Atlantic Treaty Organization October 31, 2019, https://www.nato.int/cps/en/natohq/official_texts_170408.htm.

115. Windham, "Timeline: Ten Years of Russian Cyber Attacks on Other Nations," supra.

116. Stelzenmüller, "The Impact of Russian Interference on Germany's 2017 Election," supra.

117. Windham, "Timeline: Ten Years of Russian Cyber Attacks on Other Nations," supra.

118. Stelzenmüller, "The Impact of Russian Interference on Germany's 2017 Election," supra.

119. Stefan Meister, "The 'Lisa Case': Germany as a Target of Russian Disinformation," *NATO Review*, July 25, 2016, https://www.nato.int/docu/review/articles/2016/07/25/the-lisa-case-germany-as-a-target-of-russian-disinformation/index.html.

120. Lisa-Maria N. Neudert, "Computational Propaganda in Germany: A Cautionary Tale," in Samuel Woolley and Philip N. Howard (eds.), Working Paper 2017.7 (Oxford, UK: Project on Computational Propaganda), 31. http://comprop.oii.ox.ac.uk/wp-content/uploads/sites/89/2017/06/Comprop-Germany.pdf.

121. Stelzenmüller, "The Impact of Russian Interference on Germany's 2017 Election," supra.

122. Dmitri Trenin, "Russia and Germany: From Estranged Partners to Good Neighbors," *Carnegie Moscow Center*, June 2018, https://carnegieendowment.org/files/Article_Trenin_RG_2018_Eng.pdf.

123. Oliphant, Mulholland, Huggler, and Boztas. "How Vladimir Putin and Russia Are Using Cyberattacks," supra.

124. "Successfully Countering Russian Electoral Interference," *CSIS Briefs*, June 21, 2018, https://www.csis.org/analysis/successfully-countering-russian-electoral-interference.

125. Agence France-Presse, "Norway Accuses Group Linked to Russia of Carrying out Cyber-Attack," *The Guardian*, February 3, 2017, https://www.theguardian.com/technology/2017/feb/03/norway-accuses-group-linked-to-russia-of-carrying-out-cyber-attack.

126. Bruce Sussman, "Cyber Attack Motivations: Russia vs. China," *Secure World*, June 3, 2019, https://www.secureworldexpo.com/industry-news/why-russia-hacks-why-china-hacks.

127. Russian Federation Office of the President, *The Military Doctrine of the Russian Federation* as posted by the Russian Embassy to the United Kingdom of Great Britain and Northern Ireland June 29, 2015, https://rusemb.org.uk/press/2029.

128. Ibid.

129. Ibid.

130. Louise Matsakis, "What Happens if Russia Cuts Itself Off from the Internet?" *Wired* February 12, 2019, https://www.wired.com/story/russia-internet-disconnect-what-happens/.

131. Catalin Cimpanu, "Russia to Disconnect from the Internet as Part of a Planned Test," *ZDNet* February 11, 2019, https://www.zdnet.com/article/russia-to-disconnect-from-the-internet-as-part-of-a-planned-test/.

132. Max Seddon and Henry Foy, "Russian Technology: Can the Kremlin Control the internet? *Financial Times*, June 4, 2019. https://www.ft.com/content/93be9242-85e0-11e9-a028-86cea8523dc2.

133. Marina Sadyki, "National Report on e-Commerce Development in Russia," *UN Industrial Development Organization* 2017, https://www.unido.org/api/opentext/documents/download/9920890/unido-file-9920890.

134. Katerina Mikheeva, "Why the Russian Ecommerce Market Is Worth the Hassle for Western Companies," April 9, 2019, https://www.digitalcommerce360.com/2019/04/09/why-the-russian-ecommerce-market-is-worth-the-hassle-for-western-companies/.

135. Marina Sadyki, "National Report on e-Commerce Development in Russia," supra.

136. Ibid.

137. Connell and Vogler, "Russia's Approach to Cyber Warfare," supra.

138. Tikk and Kerttunen, "Parabasis: Cyber-Diplomacy in Stalemate," supra.

139. https://www.leidensafetyandsecurityblog.nl/articles/two-incompatible-approaches-to-governing-cyberspace-hinder-global-consensus.

140. Tikk and Kerttunen, "Parabasis: Cyber-Diplomacy in Stalemate," supra.

141. Ibid.

142. Shanghai Cooperation Organization, "Agreement on Cooperation in Ensuring International Information Security between the Member States of the Shanghai Cooperation Organization," June 16, 2009 file:///C:/Users/davis/Downloads/Agreement_on_Cooperation_in_Ensuring_International_Information_Security_between_the_Member_States_of_the_SCO.pdf

143. The Ministry of Foreign Affairs of the Russian Federation, "Convention on International Information Security," September 22, 2011, https://carnegieendowment.org/files/RUSSIAN-DRAFT-CONVENTION-ON-INTERNATIONAL-INFORMATION-SECURITY.pdf.

144. CIS Information Security Agreement was signed by heads of CIS states in St. Petersburg on November 20, 2013.

145. "International code of conduct for information security," Annex to the letter dated January 9, 2015, from the Permanent Representatives of China, Kazakhstan, Kyrgyzstan, the Russian Federation, Tajikistan, and Uzbekistan to the United Nations addressed to the Secretary-General, UNGA A/69/ 723 (January 13, 2015), and UNGA A/66/359 (September 14, 2011).

146. David Ignatius, "Russia Is Pushing to Control Cyberspace. We Should All Be Worried" *The Washington Post*, October 24, 2017, https://www.washingtonpost.com/opinions/global-opinions/russia-is-pushing-to-control-cyberspace-we-should-all-be-worried/2017/10/24/7014bcc6-b8f1-11e7-be94-fabb0f1e9ffb_story.html.

147. The Soviet Pravda has two successors as decided in court—Pravda.Ru and Pravda of the Communist Party—both successors to the newspaper Pravda, established in 1912. See "There is no Pravda. There is Pravda.Ru," September 16, 2013, https://www.pravdareport.com/opinion/125664-pravda_mccain/.

147. Читайте больше на https://www.pravdareport.com/opinion/125664-pravda_mccain/.

148. Vadim Gorshenin, "Russia to Create Cyber-Warfare Units," *Pravda*, August 28, 2013, http://english.pravda.ru/russia/politics/29-08-2013/125531-cyber_warfare-0/.

149. Pasha Sharikov, "Cybersecurity in Russian-U.S. Relations," *Center for International and Security Studies* at Maryland Policy Brief April 2013.

150. Markoff and Kramer, "U.S. and Russia Differ on a Treaty for Cyberspace," supra.

151. R. I. A. Novosti and Mikhail Fomichev, "Russia to Press for International Internet Behavior Code to Fight Emerging Threats," *Russia Times*, August 1, 2013, http://rt.com/politics/russia-internet-international-code-893/.

152. Ibid.

153. Ibid.

154. US White House, "US and Russia Sign Cyber Security Pact" as quoted in *The Atlantic Council*, June 18, 2013, https://www.atlanticcouncil.org/blogs/natosource/us-and-russia-sign-cyber-security-pact.

155. Julia Joffee, "How State-Sponsored Blackmail Works in Russia," *The Atlantic*, January 11, 2017, https://www.theatlantic.com/international/archive/2017/01/kompromat-trump-dossier/512891/.

156. Alexander Dugin, *The Fourth Political Theory*. Translated by Mark Sleboda; Michael Millerman. Arktos Media, 2012.

157. "Is the Russian Orthodox Church Serving God or Putin?" *Deutsche Welt*, April 26, 2017, https://www.dw.com/en/is-the-russian-orthodox-church-serving-god-or-putin/a-38603157.

CHAPTER 4

1. Mikk Raud, "China and Cyber: Attitudes, Strategies, Organisation," *NATO Cooperative Cyber Defence Centre of Excellence*, 2016, https://ccdcoe.org/multimedia/national-cyber-security-organisation-china.html.

2. Xiaoxia, "China has 854 Million Internet Users: Report," *Xinhua*, August 30, 2019, http://www.xinhuanet.com/english/2019-08/30/c_138351278.htm.

3. Central Network Security and Informatization Leading Group, of the National Internet Information Office, *National Cyberspace Security Strategy*, December 27, 2016, http://www.cac.gov.cn/2016-12/27/c_1120195926.htm.

4. Cate Cadell, "Chinese State Media says U.S. should Take Some Blame for Cyber Attack," *Reuters*, May 13, 2017, http://www.reuters.com/article/us-cyber-attack-china-idUSKCN18D0G5.

5. Gary Robbins, "Why are China and Russia Getting Hit Hard by Cyber Attack, but not the U.S.?" *The San Diego Union-Tribune*, May 15, 2017, http://www.sandiegouniontribune.com/news/cyber-life/sd-me-ransomware-update-20170515-story.html.

6. Morse, "Snowden: Alleged NSA Attack," supra.

7. The State Council Information Office, "China's National Defense in the New Era" July 2019.

8. *National Cyberspace Security Strategy*, December 27, 2016 supra.

9. Ibid.

10. Ibid.

11. Joe Reynolds, "China's Evolving Perspectives on Network Warfare: Lessons from the Science of Military Strategy," *China Brief* 15:8, 2015, https://jamestown.org/program/chinas-evolving-perspectives-on-network-warfare-lessons-from-the-science-of-military-strategy/#.V1BM2_krK70.

12. David E. Sanger, "Chinese Curb Cyberattacks on U.S. Interests, Report Finds," *New York Times*, June 20, 2016, https://www.nytimes.com/2016/06/21/us/politics/china-us-cyber-spying.html.

13. Shannon Tiezzi, "Xi Jinping Leads China's New Internet Security Group," *The Diplomat* February 28, 2014, https://thediplomat.com/2014/02/xi-jinping-leads-chinas-new-internet-security-group/

14. State Council of the People's Republic of China, "State Council releases five-year plan on informatization," December 27, 2016, http://english.www.gov.cn/policies/latest_releases/2016/12/27/content_281475526646686.htm.

15. Ibid.

16. Ibid.

17. Raud, "China and Cyber: Attitudes, Strategies, Organisation," supra.

18. Ibid.

19. Ibid.

20. Freedom House, *Freedom on the Net, 2019: The Crisis of Social Media*, November 2019, https://freedomhouse.org/sites/default/files/2019-11/11042019_Report_FH_FOTN_2019_final_Public_Download.pdf.

21. Yan Luo, Zhijing Yu, and Nicholas Shepherd, "China's Ministry of Public Security Issues New Personal Information Protection Guideline," April 19, 2019, https://www.insideprivacy.com/data-security/chinas-ministry-of-public-security-issues-new-personal-information-protection-guideline/.

22. Raud, "China and Cyber: Attitudes, Strategies, Organisation," supra.

23. Jack Stubbs, Joseph Menn, and Christopher Bing, "China Hacked Eight Major Computer Service Firms in Years-Long Attack," *Reuters*, June 26, 2019, https://www.reuters.com/article/us-china-cyber-cloudhopper-companies-exc/exclusive-china-hacked-eight-major-computer-services-firms-in-years-long-attack-idUSKCN1TR1D4.

24. Nectar Gan, "What Do We Actually Know about China's Mysterious Spy Agency?" *South China Morning Post*, December 22, 2018, https://www.scmp.com/news/china/politics/article/2179179/what-do-we-actually-know-about-chinas-mysterious-spy-agency.

25. As quoted in Kevin Townsend, "The United States and China—A Different Kind of Cyberwar," *Security Week*, January 7, 2019, https://www.securityweek.com/united-states-and-china-different-kind-cyberwar.

26. The State Council Information Office, "China's National Defense in the New Era" July 2019.

27. Annie Kowalewski, "China's Evolving Cybersecurity Strategy," Georgetown Security Studies Review October 27, 2017, http://georgetownsecuritystudiesreview.org/2017/10/27/chinas-evolving-cybersecurity-strategy/

28. The State Council Information Office, supra.

29. Ibid., 33.

30. Lewis and Timlin, *Cybersecurity and Cyberwarfare*, supra.

31. DeWeese, Steve, Bryan Krekel, George Bakos and Christopher Barnett. US-China Economic and Security Review Commission Report on the Capability of the People's Republic of China to Conduct Cyber Warfare and Computer Network Exploitation (West Falls Church, VA: Northrop Grumman Corporation Information Systems Sector, 2009), p. 18.

32. "China's National Defense in the New Era," supra.

33. Vivien Pik-kwan Chan, "SCMP Report on PRC Officials Condemning Hacker Attacks," *South China Morning Post*, May 8, 2001.

34. Capability of the People's Republic of China to Conduct Cyber Warfare, supra., 37–38.

35. Ibid., 37.

36. Ibid., 40.

37. Reynolds, "China's Evolving Perspectives on Network Warfare," supra.

38. The State Council Information Office of the People's Republic of China, *China's Military Strategy*, May 2015, Beijing.

39. Tetsuro Kosaka, "China's Military Reorganization Could be a Force for Destabilization," *Nikkei Asian Review*, January 28, 2016, https://asia.nikkei.com/Politics/China-s-military-reorganization-could-be-a-force-for-destabilization.

40. Ibid.

41. As quoted in Townsend, "The United States and China," supra.

42. Sanger, "Chinese Curb Cyberattacks on U.S. Interests, Report Finds," supra.

43. Jim Finkle, Joseph Menn, and Aruna Viswanatha, "U.S. Accuses China of Cyber Spying on American Companies," *Reuters*, November 20, 2014, https://www.reuters.com/article/us-cybercrime-usa-china/u-s-accuses-china-of-cyber-spying-on-american-companies-idUSKCN0J42M520141120.

44. Raud, "China and Cyber: Attitudes, Strategies, Organisation," supra.

45. The State Council, The People's Republic of China, "The National Medium- and Long-Term Program for Science and Technology Development (2006–2020)" 2006, https://www.itu.int/en/ITUD/Cybersecurity/Documents/National_Strategies_Repository/China_2006.pdf

46. Ibid.

47. The originator of the strategy, now retired major general Dai Qingmin, a prolific and outspoken supporter of modernizing the military's cyberwarfare capabilities, first described the combined use of network and electronic warfare as early as 1999 in

articles and a book entitled *An Introduction to Information Warfare*, written while he was a member of faculty at the military's Electronic Engineering Academy. General Dai was promoted in 2000.

48. Steve DeWeese, Bryan Krekel, George Bakos, and Christopher Barnett, US-China Economic and Security Review Commission Report on the Capability of the People's Republic of China to Conduct Cyber Warfare and Computer Network Exploitation (Northrop Grumman Corporation's Information Systems Sector, October 9, 2009), 14–15.

49. "Countering Enemy "Informationized Operations" in War and Peace," Center for Strategic and Budgetary Assessments 2013, https://www.esd.whs.mil/Portals/54/Documents/FOID/Reading%20Room/Other/Litigation%20Release%20-%20Countering%20Enemy%20Informationized%20Operations%20in%20Peace%20and%20War.pdf.

50. Ibid., 6–7, 13.

51. Ibid.

52. Ibid.

53. *China's National Defense in 2004*, December 27, 2004, http://www.china.org.cn/e-white/20041227/index.htm

54. Ibid.

55. 32 "China's National Defense in 2004," *Information Office of the State Council of the People's Republic of China*, 2004, http://english.peopledaily.com.cn/whitepaper/defense2004/defense2004.html.

56. "The PLA's resolve to catch up with US air power is beyond doubt. . . . The PLAAF is set to argue capacity for . . . integrated information warfare, based on effective cyber warfare assets and strategic early warning and monitoring systems." See You Ji, *China's Military Transformation* (Cambridge, UK: Polity Press, 2016), 160–61.

57. Ibid., 6, 31.

58. Wang Houqing and Zhang Xingye, chief editors, *The Science of Campaigns* (National Defense University Press: Beijing, May 2000). See Chapter six, section one for an overview of information warfare in campaign settings. And Peng Guangqiang and Yao Youzhi (eds.), *The Science of Military Strategy* (Beijing: Military Science Publishing House, English edition, 2005), 336–38.

59. Capability of the People's Republic of China to Conduct Cyber Warfare supra, 6–7, 13.

60. Katherin Hille and Christian Shepard, "Taiwan: Concern Grows Over China's Invasion Threat," *Financial Times*, January 8, 2020, https://www.ft.com/content/e3462762-3080-11ea-9703-eea0cae3f0de.

61. Ibid, 20.

62. David Spencer, "Why the Risk of Chinese Cyber Attacks Could Affect Everyone in Taiwan," *Taiwan News*, July 13, 2018, https://www.taiwannews.com.tw/en/news/3481423

63. Capability of the People's Republic of China to Conduct Cyber Warfare, supra., 20.

64. Joseph Menn, "China-Based Campaign Breached Satellite, Defense Companies: Symantec," *Reuters*, June 19, 2018, https://www.reuters.com/article/us-china-usa-cyber/china-based-campaign-breached-satellite-defense-companies-symantec-idUSKBN1JF2X0.

65. Eileen Yu, "China Dispatches Online Army," *ZDNet*, May 27, 2011, http://www.zdnet.com/china-dispatches-online-army-2062300502/.

66. Amy Chang, "China's Maodun: A Free Internet Caged by the Chinese Communist Party," *China Brief* XV:8, 2015.

67. Second Bureau of the People's Liberation Army (PLA) General Staff Department's (GSD) 3rd Department (总参三部二局), which is most commonly known by its Military Unit Cover Designator (MUCD) as Unit 61398 (61398部队).

68. Mandiant, *APT1: Exposing One of China's Cyber Espionage Units* Mandiant Report (2013).

69. Ibid., 20.

70. Sanger, "Chinese Curb Cyberattacks on U.S. Interests, Report Finds," supra.

71. "China Media: US Ambassador Gary Locke's Legacy," *BBC News*, November 21, 2013, http://www.bbc.co.uk/news/world-asia-china-25029646

72. "Beijing's Cyberspies Step Up Surveillance of Ethnic Groups with New Language-Tracking Technology," *South China Morning Post*, November 20, 2013, http://www.scmp.com/news/china/article/1361547/central-government-cyberspies-step-surveillance-ethnic-groups-new.

73. James Glanz and John Markoff, "Vast Hacking by a China Fearful of the Web," *The New York Times*, December 4, 2010, http://www.nytimes.com/2010/12/05/world/asia/05wikileaks-china.html?pagewanted=all_r=1.

74. Paul Mozur, "One Month, 500,000 Face Scans: How China Is Using A.I. to Profile a Minority," *The New York Times*, May 5, 2019, https://www.nytimes.com/2019/04/14/technology/china-surveillance-artificial-intelligence-racial-profiling.html?action=clickmodule=RelatedLinkspgtype=Article.

75. The State Council Information Office, supra.

76. Sanger, "Chinese Curb Cyberattacks on U.S. Interests, Report Finds," supra.

77. Ibid.

78. Elsa B. Kania, "Made in China 2025, Explained," *The Diplomat*, February 1, 2019, https://thediplomat.com/2019/02/made-in-china-2025-explained/

79. Mandiant, *APT1: Exposing One of China's Cyber Espionage Units*, supra.

80. Second Bureau of the People's Liberation Army (PLA) General Staff Department's (GSD) 3rd Department (总参三部二局), which is most commonly known by its Military Unit Cover Designator (MUCD) as Unit 61398 (61398部队).

81. Unit 61398 is partially situated on Datong Road (大同路) in Gaoqiaozhen (高桥镇), which is located in the Pudong New Area (浦东新区) of Shanghai (上海).

82. Mandiant, *APT1: Exposing One of China's Cyber Espionage Units* supra.

83. Ibid.

84. Brendan I. Koerner, "Inside the Cyberattack That Shocked the US Government," *Wired*, October 13, 2016, https://www.wired.com/2016/10/inside-cyberattack-shocked-us-government/

85. Kristie Lu Stout, "Cyber Warfare: Who is China Hacking Now?" *CNN* September 29, 2016, http://www.cnn.com/2016/09/29/asia/china-cyber-spies-hacking/index.html

86. Freedom House, *Freedom on the Net, 2019*, supra.

87. Ibid.

88. China's State Council, "Planning Outline for the Construction of a Social Credit System (2014–2020)," April 25, 2015, https://chinacopyrightandmedia.wordpress.com/2014/06/14/planning-outline-for-the-construction-of-a-social-credit-system-2014-2020/.

89. "Planning Outline for the Construction of a Social Credit System (2014–2020)," supra.

90. Ibid.

91. Jamie Horsley, "China's Orwellian Social Credit Score Isn't Real," *Foreign Policy*, November 16, 2018, https://foreignpolicy.com/2018/11/16/chinas-orwellian-social-credit-score-isnt-real/

92. Alexander Chipman Koty, "China's Social Credit System: COVID-19 Triggers Some Exemptions, Obligations for Businesses," *China Briefing* March 26, 2020, https://www.china-briefing.com/news/chinas-social-credit-system-covid-19-triggers-some-exemptions-obligations-businesses/.

93. Ibid.

94. Genia Kostka, "What Do People in China Think About 'Social Credit' Monitoring?" *The Washington Post*, March 21, 2019, https://www.washingtonpost.com/politics/2019/03/21/what-do-people-china-think-about-social-credit-monitoring/?utm_term=.49fe491dd67b.

95. Rachel Botsman, "Big Data Meets Big Brother as China Moves to Rate its Citizens," *Wired*, October 21, 2017, https://www.wired.co.uk/article/chinese-government-social-credit-score-privacy-invasion.

96. Simina Mistreanu, "Life Inside China's Social Credit Laboratory," April 3, 2018 *Foreign Policy*. https://foreignpolicy.com/2018/04/03/life-inside-chinas-social-credit-laboratory/.

97. Kostka, "What Do People in China Think About 'Social Credit' Monitoring?" supra.

98. Ibid.

99. Horsley, "China's Orwellian Social Credit Score Isn't Real," supra.

100. "Planning Outline for the Construction of a Social Credit System (2014–2020)," supra.

101. Abishur Prakash, "Facial Recognition Cameras and AI: 5 Countries with the Fastest Adoption," *Robotics Business Review*, December 21, 2018, https://www.roboticsbusinessreview.com/ai/facial-recognition-cameras-5-countries/.

102. Ibid.

103. Bethany Allen-Ebrahimian, "China Cables Exposed: China's Operating Manuals for Mass Internment and Arrest by Algorithm" *International Consortium of Investigative Journalists*, November 24, 2019, https://www.icij.org/investigations/china-cables/exposed-chinas-operating-manuals-for-mass-internment-and-arrest-by-algorithm/.

104. Nicole Kobie, "The Complicated Truth About China's Social Credit System," *Wired*, January 21, 2019, https://www.wired.co.uk/article/china-social-credit-system-explained.

105. Mistreanu, "Life Inside China's Social Credit Laboratory," supra.

106. Ibid.

107. Stepan Kravchenko, "Russia More Prey Than Predator to Cyber Firm Wary of China," *Bloomberg News*, August 25, 2016, https://www.bloomberg.com/news/articles/2016-08-25/russia-more-prey-than-predator-to-cyber-firm-wary-of-china.

108. Ibid.

109. Ibid.

110. In January 2016, Mr. Trump said he would favor a 45 percent tariff on Chinese exports to the United States, proposing the idea during a wide-ranging meeting with members of the editorial board of *The New York Times*. See Maggie Haberman, "Donald Trump Says He Favors Big Tariffs on Chinese Exports," *The New York Times,* January 7, 2016, https://www.nytimes.com/politics/first-draft/2016/01/07/donald-trump-says-he-favors-big-tariffs-on-chinese-exports/.

111. Jonathan Cheng and Josh Chin, "China Hacked South Korea Over Missile Defense, U.S. Firm Says," *Wall Street Journal,* April 21, 2017, https://www.wsj.com/articles/chinas-secret-weapon-in-south-korea-missile-fight-hackers-1492766403?emailToken=JRrydPtyYnqTg9EyZsw31FwuZ7JNEOKCXF7LaW/HM1DLsjnUp6e6wLgph560pnmiTAN/5ssf7moyADPQj2p2Gc+YkL1yi0zhIiUM9M6aj1HTYQ==

112. Sean Gallagher, "Researchers Claim China Trying to Hack South Korean Missile Defense Efforts," *ARS Technica*, April 21, 2017, https://arstechnica.com/security/2017/04/researchers-claim-china-trying-to-hack-south-korea-missile-defense-efforts/.

113. "China Blamed After ASIO Blueprints Stolen in Major Cyber Attack on Canberra HQ," *Australian Broadcasting Corporation News*, May 27, 2013, http://www.abc.net.au/news/2013-05-27/asio-blueprints-stolen-in-major-hacking-operation/4715960

114. Greg Weston, "Foreign Hackers Attack Canadian Government: Computer Systems at 3 Key Departments Penetrated," *Canadian Broadcasting Corporation News*, February 16, 2011, http://www.cbc.ca/news/politics/foreign-hackers-attack-canadian-government-1.982618

115. David E. Sanger, Davis Barboza, and Nicole Perlroth, "Chinese Army Unit Is Seen as Tied to Hacking Against U.S.," *The New York Times*, February 18, 2013, https://www.nytimes.com/2013/02/19/technology/chinas-army-is-seen-as-tied-to-hacking-against-us.html.

116. Rosemary Barton, "Chinese Cyberattack Hits Canada's National Research Council," *CBC News*, July 29, 2014, http://www.cbc.ca/news/politics/chinese-cyberattack-hits-canada-s-national-research-council-1.2721241.

117. Yatish Yadav, "80,000 Cyberattacks on December 9 and 12 After Note Ban," *New India Express*, December 19, 2016, http://www.newindianexpress.com/nation/2016/dec/19/80000-cyber-attacks-on-december-9-and-12-after-note-ban-1550803.html.

118. Naveen Goud, "China Cyberattacks Indian SUKHOI 30 Jet Fighters!" *Cybersecurity Insiders* June 5, 2017, https://www.cybersecurity-insiders.com/china-cyber-attacks-indian-sukhoi-30-jet-fighters/

119. Drogin, "Russians Seem to Be Hacking into Pentagon," supra.

120. Bradley Graham, "Hackers Attack Via Chinese Web Sites," *The Washington Post,* August 25, 2005, http://www.washingtonpost.com/wp-dyn/content/article/2005/08/24/AR2005082402318.html.

121. Sanger, Barboza, and Perlroth, "Chinese Army Unit Is Seen as Tied to Hacking Against U.S.," supra.

122. Ibid.

123. Ibid.

124. Chauncey Jung, "China Ventures Outside the Great Firewall, Only to Hit the Brick Wall of Online Etiquette and Trolls," *South China Morning,* Post February 23, 2020, https://www.scmp.com/comment/opinion/article/3051757/china-ventures-outside-great-firewall-only-hit-brick-wall-online.

125. Ibid.

126. Ibid.

127. Sussman, "Cyber Attack Motivations: Russia vs. China," supra.

128. "China Condemns Decision by Google to Lift Censorship," *BBC,* March 23, 2010, http://news.bbc.co.uk/2/hi/asia-pacific/8582233.stm.

129. Glanz and Markoff, "Vast Hacking by a China Fearful of the Web," supra.

130. Sue-Lin Wong and Michael Martina, "China Adopts Cyber Security Law in Face of Overseas Opposition," *Reuters Technology News*, November 7, 2016, http://www.reuters.com/article/us-china-parliament-cyber-idUSKBN132049

131. As quoted in Ibid.

132. Kevin Wei Wang, Jonathan Woetzel, Jeongmin Seong, James Manyika, Michael Chui, and Wendy Wong, "Digital China: Powering the Economy to global Competitiveness," *McKinsey Global Institute*, December 2017, https://www.mckinsey.com/featured-insights/china/digital-china-powering-the-economy-to-global-competitiveness.

133. Zen Soo, "Here's How China's New e-Commerce Law Will Affect Consumers, Platform Operators," *South China Morning Post,* January 1, 2019, https://www.scmp.com/tech/apps-social/article/2180194/heres-how-chinas-new-e-commerce-law-will-affect-consumers-platform.

134. Tikk and Kerttunen, "Parabasis: Cyber-diplomacy in Stalemate," supra.

135. An International Code of Conduct for Information Security: China's perspective on building a peaceful, secure, open, and cooperative cyberspace. Remarks delivered on February 10, 2014 at UNIDIR: "Nowadays, the information 'highway' has reached almost every corner of the world. It is of great concern, however, that in this virtual space where traffic is very heavy, there is still no comprehensive 'traffic rules.' As a result, 'traffic accidents' in information and cyber space constantly occur with ever increasing damage and impact."

136. Tikk and Kerttunen, "Parabasis: Cyber-Diplomacy in Stalemate," supra.

137. Ibid.

138. Ku, "Forcing China to Accept that International Law Restricts Cyber Warfare," supra.

139. Tikk and Kerttunen, "Parabasis: Cyber-diplomacy in Stalemate," supra.

140. Adam Segal, "How China is Preparing for Cyberwar," *Christian Science Monitor*, March 20, 2017, http://www.csmonitor.com/World/Passcode/Passcode-Voices/2017/0320/How-China-is-preparing-for-cyberwar.

141. Tikk and Kerttunen, "Parabasis: Cyber-diplomacy in Stalemate," supra.

142. Glanz and Markoff, "Vast Hacking by a China Fearful of the Web," supra.

143. Tzu, *The Art of War*, supra., 1963.

144. Wong and Martina, "China Adopts Cyber Security Law in Face of Overseas Opposition," supra.

CHAPTER 5

1. Reuters, "Trump Says Discussed Forming Cyber Security Unit with Putin," *Reuters*, July 9, 2017, https://www.reuters.com/article/us-g20-germany-putin-trump/trump-says-discussed-forming-cyber-security-unit-with-putin-idUSKBN19U0HX.

2. "NATO Staying Strong in Cyberspace," *NATO Headquarters News*, March 24, 2020, https://www.nato.int/cps/en/natohq/news_174499.htm?selectedLocale=en.

3. "Strategic Arms Limitation Talks (SALT I)," *The Nuclear Threat Initiative*, October 26, 2011, https://www.nti.org/learn/treaties-and-regimes/strategic-arms-limitation-talks-salt-i-salt-ii/

4. Tyler Quinn, "The Bear's Side of the Story: Russian Political and Information Warfare," *RealClearDefense*, June 27, 2018, https://www.realcleardefense.com/articles/2018/06/27/the_bears_side_of_the_story_russian_political_and_information_warfare_113564.html

5. Ariana Rowberry, "Sixty Years of 'Atoms for Peace' and Iran's Nuclear Program," *The Brookings Institute*, December 18, 2013, https://www.brookings.edu/blog/up-front/2013/12/18/sixty-years-of-atoms-for-peace-and-irans-nuclear-program/.

6. Dwight D. Eisenhower, "Atoms for Peace Speech," *Address by Mr. Dwight D. Eisenhower, President of the United States of America, to the 470th Plenary Meeting of the United Nations General Assembly*, December 8, 1953, https://www.iaea.org/about/history/atoms-for-peace-speech.

7. Sergey Lavrov, "Remarks by Foreign Minister Sergey Lavrov at the XXII Assembly of the Council on Foreign and Defence Policy," *Valdai Discussion Club*, November 2, 2014, https://valdaiclub.com/a/highlights/remarks_by_foreign_minister_sergey_lavrov_at_the_xxii_assembly_of_the_council_on_foreign_and_defence/.

8. US President George Washington, "From George Washington to John Trumbull, 25 June 1799," *National Archives*. June 25, 1799, https://founders.archives.gov/documents/Washington/06-04-02-0120

9. Bryon Tau, "Government Tracking How People Move Around in Coronavirus Pandemic," *The Wall Street Journal*, March 28, 2020, https://www.wsj.com/articles/government-tracking-how-people-move-around-in-coronavirus-pandemic-11585393202.

10. Andrei Soldatov and Irina Borogan, *The Red Web: The Struggle Between Russia's Digital Dictators and the New Online Revolutionaries* (New York: Hachette Book Group, 2017)

11. Ibid.

12. Editorial Board, "The Coronavirus Gives Russia and China Another Opportunity to Spread Their Disinformation," *The Washington Post*, March 29, 2020, https://www.washingtonpost.com/opinions/the-coronavirus-gives-russia-and-china-another-opportunity-to-spread-their-disinformation/2020/03/29/8423a0f8-6d4c-11ea-a3ec-70d7479d83f0_story.html.

13. Ibid.

14. Mary Ilyushina, "How Russia Is Using Authoritarian Tech to Curb Coronavirus," *CNN*, March 29, 2020, https://www.cnn.com/2020/03/29/europe/russia-coronavirus-authoritarian-tech-intl/index.html

15. Maréchal, "Are You Upset About Russia Interfering With Elections?" supra.

16. ICANN, "What is Policy?" *Internet Corporation for Assigned Names and Numbers* https://www.icann.org/policy#what_is_policy.

17. Maréchal, "Are You Upset About Russia Interfering With Elections?" supra.

18. Chris C. Demchak and Yuval Shavitt, "China's Maxim—Leave No Access Point Unexploited: The Hidden Story of China Telecom's BGP Hijacking," *Military Cyber Affairs*, 3:1, Article 7, 2018, https://www.doi.org/https://doi.org/10.5038/2378-0789.3.1.1050.

19. Frank Bajak, "Internet Traffic Hijack Disrupts Google Services," *AP* November 12, 2018, https://apnews.com/4e2cb39354ce4e338f3ee50446597ef5

20. Laura Rosenberger, "China's Coronavirus Information Offensive: Beijing Is Using New Methods to Spin the Pandemic to Its Advantage," *Foreign Affairs*, April 22, 2020, https://www.foreignaffairs.com/articles/china/2020-04-22/chinas-coronavirus-information-offensive.

21. Paul Mozur, Raymond Zhong, and Aaron Krolik, "In Coronavirus Fight, China Gives Citizens a Color Code, With Red Flags," *The New York Times*, March 1, 2020, https://www.nytimes.com/2020/03/01/business/china-coronavirus-surveillance.html?_ga=2.231721687.974084075.1585035405-293092624.1585035405.

22. Xinhua News Agency, "I Have Seen Hangzhou at Four in the Morning, Just for an Early 'No Code'-The Recovery Record Behind the 'Health Code,'" *Xinhua*, February 21, 2020, https://baijiahao.baidu.com/s?id=1659147516916399925wfr=spiderfor=pc

23. Wang, Woetzel et al, "Digital China: Powering the Economy to Global Competitiveness," supra.

24. Tikk and Kerttunen, "Parabasis: Cyber-diplomacy in Stalemate," supra.

25. Segal, "How China Is Preparing for cyberwar," supra.

26. John D. Negroponte and Samuel J. Palmisano, "Defending an Open, Global, Secure, and Resilient Internet," *The Council on Foreign Relations,* June 2013, https://www.cfr.org/content/publications/attachments/TFR70_cyber_policy.pdf.pdf.

27. Joseph R. DeTrani, "Cyberspace: A Global Threat to Peace," *Asia Times*, October 28, 2013, http://www.atimes.com/atimes/World/WOR-02-281013.html

28. United Nations General Assembly, "International Code of Conduct for Information Security," *Report of the United Nations General Assembly*, 2015.

29. For instance, the declaration from the Lisbon Summit, which took place in November 2010 as part of a NATO initiative, said that establishing "an EU-US Working Group on Cyber-security and Cyber-crime . . . will address a number of

specific priority areas. Lisbon Summit Declaration: Issued by the Heads of State and Government participating in the meeting of the North Atlantic Council in Lisbon" North Atlantic Treaty Organization November 20, 2010, https://www.nato.int/cps/en/natolive/official_texts_68828.htm.

30. Tom Gjelten, "Seeing the Internet as an 'Information Weapon,'" *National Public Radio* September 23, 2010, http://www.npr.org/templates/story/story.php?storyId=130052701.

31. Masters, "Confronting the Cyber Threat," supra.

32. The five questions are as follows: (1) Can protected critical humanitarian infrastructure entities be "detangled" from non-protected entities in cyberspace? (2) Just as a Red Cross designates a protected entity in the physical world, is it feasible to use special markers to designate protected zones in cyberspace? (3) Should we reinterpret convention principles in light of the fact that cyberoperatives are often non-state actors? (4) Are certain cyberweapons analogous to weapons banned by the Geneva Protocol? And (5) Given the difficulties in coming up with an agreed definition for cyber war, should there be a third, "other-than-war" mode for cyberspace? See Karl Rauscher and Andrey Korotkov, *Working Towards Rules for Governing Cyber Conflict: Rendering the Geneva and Hague Conventions in Cyberspace* (First Joint Russian-U.S. report on Cyber Conflict) (Honolulu, HI: East West Center, 2011).

33. Raud, "China and Cyber: Attitudes, Strategies, Organisation," supra.

347. Gary Brown and Christopher D. Yung, "Evaluating the US-China Cybersecurity Agreement, Part 1: The US Approach to Cyberspace, How Washington approaches cyberspace and its 2015 cybersecurity agreement with China," *The Diplomat*, January 19, 2017, https://thediplomat.com/2017/01/evaluating-the-us-china-cybersecurity-agreement-part-1-the-us-approach-to-cyberspace/.

35. Central Network Security and Informatization Leading Group, of the National Internet Information Office, *National Cyberspace Security Strategy*, December 27, 2016, http://www.cac.gov.cn/2016-12/27/c_1120195926.htm.

36. Iain Sutherland, Konstantinos Xynos, Andrew Jones and Andrew Blyth, "The Geneva Conventions and Cyber-Warfare: A Technical Approach," *The RUSI Journal*, 30–39, 2015.

37. Sutherland, Xynos, Jones, and Blyth, "The Geneva Conventions and Cyber-Warfare: A Technical Approach," supra.

38. Ashley Deeks, "Tallinn 2.0 and a Chinese View on the Tallinn Process," *Lawfare*, May 31, 2015, https://www.lawfareblog.com/tallinn-20-and-chinese-view-tallinn-process.

Bibliography

"Beijing's Cyberspies Step Up Surveillance of Ethnic Groups with New Language-Tracking Technology." *South China Morning Post*, November 20, 2013. http://www.scmp.com/news/china/article/1361547/central-government-cyberspies-step-surveillance-ethnic-groups-new.

"Big Surge in Cyberattacks on Russia Amid US Hacking Hysteria—Russian Security Chief." *Russia Times*, January 15, 2017. https://www.rt.com/news/373764-surge-hacking-attacks-russia/.

"China Blamed after ASIO Blueprints Stolen in Major Cyber Attack on Canberra HQ." *Australian Broadcasting Corporation News*, May 27, 2013. http://www.abc.net.au/news/2013-05-27/asio-blueprints-stolen-in-major-hacking-operation/4715960.

"China Media: US Ambassador Gary Locke's Legacy." *BBC News*, November 21, 2013. http://www.bbc.co.uk/news/world-asia-china-25029646.

"Clarke: More Defense Needed in Cyberspace." *HometownAnnapolis.com*, Sept. 24, 2010. https://www.whitehouse.gov/the-press-office/2017/05/11/presidential-executive-order-strengthening-cybersecurity-federal.

"Countering Enemy "Informationized Operations" in War and Peace." *Center for Strategic and Budgetary Assessments*, 2013. https://www.esd.whs.mil/Portals/54/Documents/FOID/Reading%20Room/Other/Litigation%20Release%20-%20Countering%20Enemy%20Informationized%20Operations%20in%20Peace%20and%20War.pdf.

"Cyber Mission Force Achieves Full Operational Capability." *U.S. Cyber Command News Release*, May 17, 2018. https://www.defense.gov/News/Article/Article/1524747/cyber-mission-force-achieves-full-operational-capability/.

"Cyber Wars." *Agentura.Ru*, August 21, 1997. www.agentura.ru/english/equipment/.

"Deterring Hybrid Warfare." *NATO Review*, 2014. http://www.nato.int/docu/review/2014/Also-in-2014/Deterring-hybrid-warfare/EN/index.htm.

"Dutch Intelligence First to Alert U.S. about Russian Hack of Democratic Party." *Nieuwsuur*, January 25, 2018. https://nos.nl/nieuwsuur/artikel/2213767-dutch-intelligence-first-to-alert-u-s-about-russian-hack-of-democratic-party.html.

"Election Watchdog Golos Demands to be Removed from 'Foreign Agents' List after Court Victory." *Russian Times*, September 9, 2014. https://www.rt.com/politics/186452-golos-watchdog-ngo-court/.

"Estonia Hit by 'Moscow Cyber War'." *BBC News*, May 17, 2007. http://news.bbc.co.uk/go/pr/fr/-/2/hi/europe/6665145.stm.

"Flame Malware Makers Send 'Suicide' Code." *BBC News*, June 2012. http://www.bbc.co.uk/news/technology-18365844.

"Flame Virus 'Created by US and Israel as Part of Intensifying Cyber Warfare'." *The Telegraph*, June 6, 2014.

"Hackers Gain Entry into US, European Energy Sector, Symantec Warns." *Reuters*, September 6, 2017. https://www.cnbc.com/2017/09/06/hackers-gain-entry-into-us-european-energy-sector-symantec-warns.html.

"Huawei Accuses US of Cyber-Attacks and Threats to Staff." *BBC News*, September 4, 2019. https://www.bbc.com/news/business-49574890.

"International Code of Conduct for Information Security." *UNGA* A/69/ 723, January 13, 2015 and UNGA A/66/359, September 14, 2011.

"Is the Russian Orthodox Church Serving God or Putin?" *Deutsche Welt*, April 26, 2017. https://www.dw.com/en/is-the-russian-orthodox-church-serving-god-or-putin/a-38603157.

"Media Statement Regarding Reported US DoJ Probes into Huawei." *Huawei Press Release*, September 2, 2019. https://www.huawei.com/en/facts/voices-of-huawei/media-statement-regarding-reported-us-doj-probes-into-huawei.

"N. Korea's Cyber Warfare Unit in Spotlight after Attack on S. Korean Bank." *Yonhap News Agency*, May 3, 2011. http://english.yonhapnews.co.kr/national/2011/05/03/78/0301000000AEN20110503010600315F.HTML.

"Natural Disasters Cost US a Record $306 Billion Last Year." *CBS*, January 8, 2018. https://www.cbsnews.com/news/us-record-306-billion-natural-disasters-last-year-hurricanes-wilidfires/.

"Report of the International Agora 'Freedom of the Internet 2018: delegation of repression'." *Agora*, May 2, 2019. https://www.agora.legal/articles/Doklad-Mezhdunarodnoi-Agory-%C2%ABSvoboda-interneta-2018-delegirovanie-repressiy%C2%BB/18.

"Russia Enacts 'Draconian' Law for Bloggers and Online Media." *BBC News*, August 1, 2014. https://www.bbc.com/news/technology-28583669.

"Russia Internet: Law Introducing New Controls Comes into Force." *BBC News*, November 1, 2019. https://www.bbc.com/news/world-europe-50259597.

"Russian Computer Programmer Arrested in Spain: Embassy." *Reuters*, April 9, 2017. http://www.reuters.com/article/us-spain-russia-idUSKBN17B0O2.

"Statement of the NATO-Ukraine Commission." *North Atlantic Treaty Organization*, October 31, 2019. https://www.nato.int/cps/en/natohq/official_texts_170408.htm.

"Successfully Countering Russian Electoral Interference." *CSIS Briefs*, June 21, 2018. https://www.csis.org/analysis/successfully-countering-russian-electoral-interference.

"WannaCry: Are You Safe?" *Kaspersky Company*, May 15, 2017. https://www.kaspersky.co.uk/blog/wannacry-ransomware/8700/.

"Zimbabwe's Begging Bowl: Bailing Out Bandits." *The Economist*, July 9, 2016. https://www.economist.com/middle-east-and-africa/2016/07/09/bailing-out-bandits.

"Военная доктрина Российской Федерации." *Russian Presidential Executive Office*, February 5, 2010. http://news.kremlin.ru/ref_notes/461.

Agence France Presse. "Norway Accuses Group Linked to Russia of Carrying Out Cyber-Attack: Norwegian Intelligence Service PST Among Targets of Malicious Emails Believed to Have Been Sent by APT 29." *The Guardian*, February 3, 2017. https://www.theguardian.com/technology/2017/feb/03/norway-accuses-group-linked-to-russia-of-carrying-out-cyber-attack.

Alexander, Keith B. "Warfighting in Cyberspace." *Joint Forces Quarterly*, July 31, 2007. http://www.military.com/forums/0,15240,143898,00.html.

Allhoff, Fritz, et al. *Binary Bullets: The Ethics of Cyberwarfare*. New York: Oxford University Press, 2016.

Andress, Jason and Steve Winterfeld. *Cyber Warfare: Techniques, Tactics and Tools for Security Practitioners*. Rockland, MA: Syngress, 2011.

Ashford, Warwick. "Security Research Links Russia to US Election Cyberattacks: Security Researchers say the Hacking of the US Democratic National Convention's Email System is Linked to a Wider Russian Cyber Campaign." *Computer Weekly*, January 6, 2017. http://www.computerweekly.com/news/450410516/Security-research-links-Russia-to-US-election-cyber-attacks.

Baker, Prentiss O. LTC. "Psychological Operations Within the Cyber Domain." Maxwell Paper No. 52, *The Air War College*, February 17, 2010. http://www.au.af.mil/au/awc/awcgate/maxwell/mp52.pdf

Barabanov, Mikhail. "Testing a "New Look'." *Russia in Global Affairs*, December 18, 2014. https://eng.globalaffairs.ru/number/Testing-a-New-Look-17213.

Baram, Gil. "The Theft and Reuse of Advanced Offensive Cyber Weapons Pose A Growing Threat." *Council on Foreign Relations*, June 19, 2018. https://www.cfr.org/blog/theft-and-reuse-advanced-offensive-cyber-weapons-pose-growing-threat.

Barnes, Julian E. and Thomas Gibbons-Neff. "U.S. Carried Out Cyberattacks on Iran." *The New York Times*, June 22, 2019. https://www.nytimes.com/2019/06/22/us/politics/us-iran-cyber-attacks.html.

Barrett, Brian. "For Russia, Unravelling US Democracy Was Just Another Day Job." *Wired*, February 17, 2018. https://www.wired.com/story/mueller-indictment-internet-research-agency/

Barton, Rosemary. "Chinese Cyberattack Hits Canada's National Research Council." *CBC News*, July 29, 2014. http://www.cbc.ca/news/politics/chinese-cyberattack-hits-canada-s-national-research-council-1.2721241

Bencsáth, Boldizsár, Gábor Pék, Levente Buttyán, and Márk Félegyházi. "Duqu: A Stuxnet-Like Malware Found in the Wild." Laboratory of Cryptography and System Security, Budapest University of Technology and Economics' Department of Telecommunications, October 2011. http://www.crysys.hu/publications/files/bencsathPBF11duqu.pdf.

Bērziņš, Jānis. "Russia's New Generation Warfare in Ukraine." National Defence Academy of Latvia Center for Security and Strategic Research Policy, Paper No. 2, April 2014. http://www.sldinfo.com/wp-content/uploads/2014/05/New-Generation-Warfare.pdf.

Beuth, Von Patrick, Kai Biermann, Martin Klingst und Holger Stark. "Cyberattack on the Bundestag: Merkel and the Fancy Bear." *Zeit Online*, May 12, 2017. https://www.zeit.de/digital/2017-05/cyberattack-bundestag-angela-merkel-fancy-bear-hacker-russia.

Botsman, Rachel. "Big Data Meets Big Brother as China Moves to Rate Its Citizens." *Wired*, October 21, 2017. https://www.wired.co.uk/article/chinese-government-social-credit-score-privacy-invasion.

Bradshaw, Samantha and Philip N. Howard. "Troops, Trolls and Troublemakers: A Global Inventory of Organized Social Media Manipulation." In *Project on Computational Propaganda*, edited by Samuel Woolley and Philip N. Howard. Oxford: Oxford University, 2017. http://comprop.oii.ox.ac.uk/research/troops-trolls-and-trouble-makers-a-global-inventory-of-organized-social-media-manipulation/.

Bredemeier, Ken. "Russia Demands Return of 2 Shuttered Compounds in US." *Voice of America*, July 17, 2017. https://www.voanews.com/usa/russia-demands-return-2-shuttered-compounds-us.

Brown, Gary and Christopher D. Yung. "Evaluating the US-China Cybersecurity Agreement, Part 1: The US Approach to Cyberspace, How Washington approaches cyberspace and its 2015 cybersecurity agreement with China." *The Diplomat*, January 19, 2017. https://thediplomat.com/2017/01/evaluating-the-us-china-cybersecurity-agreement-part-1-the-us-approach-to-cyberspace/.

Cadell, Cate. "Chinese State Media Says U.S. Should Take Some Blame for Cyberattack." *Reuters*, May 13, 2017. http://www.reuters.com/article/us-cyber-attack-china-idUSKCN18D0G5.

Cadwalladr, Carole. "Fresh Cambridge Analytica Leak 'Shows Global Manipulation is Out of Control'." *The Guardian*, January 4, 2020. https://www.theguardian.com/uk-news/2020/jan/04/cambridge-analytica-data-leak-global-election-manipulation.

Carlin, John P. *Dawn of the Code War: America's Battle Against Russia, China, and the Rising Global Cyber Threat.* New York: Hatchett Book Group, 2018.

Carr, Jeffery. *Inside Cyber Warfare: Mapping the Cyber Underworld.* Sebastopol, CA: O'Reilly Media, 2009.

Center for Internet Security. "Critical Security Controls for Effective Cyber Defense." *Center for Internet Security*, October 15, 2015. https://web.archive.org/web/20160809003039/https://www.cisecurity.org/critical-controls/documents/CSC-MASTER-VER%206.0%20CIS%20Critical%20Security%20Controls%2010.15.2015.pdf.

Central Intelligence Agency, "Studies in Intelligence: A Collection of Articles on the Historical, Operational, Doctrinal, and Theoretical Aspects of Intelligence." *National Security Archive*, Winter 1999–2000. https://cryptome.org/nsa-shamrock.htm.

"Central Network Security and Informatization Leading Group of the National Internet Information Office," *National Cyberspace Security Strategy*. December 27, 2016. http://www.cac.gov.cn/2016-12/27/c_1120195926.htm.

Chan, Vivien Pik-kwan. "SCMP Report on PRC Officials Condemning Hacker Attacks." *South China Morning Post*, May 8, 2001.

Chang, Amy. "China's Maodun: A Free Internet Caged by the Chinese Communist Party." *China Brief*, Volume XV, Issue 8, April 17, 2015.

Chapple, Amos. "The Art Of War: Russian Propaganda In WWI." *Radio Free Europe/Radio Liberty*, 2018. https://www.rferl.org/a/russias-world-war-one-propaganda-posters/29292228.html.

Chen, Adrian. "The Agency." *The New York Times*, June 2, 2015. https://www.nytimes.com/2015/06/07/magazine/the-agency.html.

Cheng, Jonathan and Josh Chin. "China Hacked South Korea Over Missile Defense, U.S. Firm Says." *The Wall Street Journal*, April 21, 2017. https://www.wsj.com/articles/chinas-secret-weapon-in-south-korea-missile-fight-hackers-1492766403?emailToken=JRrydPtyYnqTg9EyZsw31FwuZ7JNEOKCXF7LaW/HM1DLsjnUp6e6wLgph560pnmiTAN/5ssf7moyADPQj2p2Gc+YkL1yi0zhIiUM9M6aj1HTYQ==.

China Internet Network Information Center. "The 41st "Statistical Report on the Development of China's Internet Network." *China Internet Network Information Center*, January 31, 2018. http://cnnic.cn/gywm/xwzx/rdxw/201801/t20180131_70188.htm.

China, the People's Republic of, Information Office of the State Council. "China's National Defense in 2004." December 27, 2004. http://www.china.org.cn/e-white/20041227/index.htm.

China, the People's Republic of, Information Office of the State Council. "Planning Outline for the Construction of a Social Credit System (2014–2020)." April 25, 2015. https://chinacopyrightandmedia.wordpress.com/2014/06/14/planning-outline-for-the-construction-of-a-social-credit-system-2014-2020/

China, the People's Republic of China, The State Council Information Office. "State Council Releases Five-Year Plan on Informatization." December 27, 2016. http://english.www.gov.cn/policies/latest_releases/2016/12/27/content_281475526646686.htm

China, the People's Republic of China, The State Council Information Office. *China's Military Strategy*. Beijing: The State Council Information Office, May 2015.

China, the People's Republic of China, The State Council Information Office, "The National Medium- and Long-Term Program for Science and Technology Development (2006–2020)." 2006. https://www.itu.int/en/ITU-D/Cybersecurity/Documents/National_Strategies_Repository/China_2006.pdf.

China, the People's Republic of China, The State Council Information Office, "China's National Defense in the New Era." July 2019. http://www.xinhuanet.com/english/2019-07/24/c_138253389.htm.

Cimpanu, Catalin. "Russia to Disconnect from the Internet as Part of a Planned Test." *ZDNet*, February 11, 2019. https://www.zdnet.com/article/russia-to-disconnect-from-the-internet-as-part-of-a-planned-test/.

Clapper, James R. "Statement for the Record: Worldwide Cyber Threats." *House Permanent Select Committee on Intelligence*, September 10, 2015. https://www.dni.gov/index.php/newsroom/testimonies/209-congressional-testimonies- 2015/1251-dni-clapper-statement-for-the-record,-worldwide-cyber-threats-before-the-housepermanent-select-committee-on-intelligence.

Clarke, Richard A. and Robert K. Knake, *Cyber War: The Next Threat to National Security and What to Do about It*. New York: Harper Collins Books, 2010.

Clayton, Mark. "More Telltale Signs of Cyber Spying and Cyberattacks Arise in Middle East." *The Christian Science Monitor*, August 21, 2012. http://www.csmonitor.com/USA/2012/0821/More-telltale-signs-of-cyber-spying-and-cyber-attacks-arise-in-Middle-East-video.

Clinton, William. *National Security Strategy* Washington, DC: Government Printing Office, 1995.

Cohen, Rouven. "New Massive Cyber-Attack an 'Industrial Vacuum Cleaner for Sensitive Information'," *Forbes*, May 28, 2012.

Cole, Matthew, Richard Esposito, Sam Biddle, and Ryan Grim, "Top-Secret NSA Report Details Russian Hacking Effort Days Before 2016 Election." *The Intercept*, June 5, 2017. https://theintercept.com/2017/06/05/top-secret-nsa-report-details-russian-hacking-effort-days-before-2016-election/.

Coll, Steve. "The Rewards (and Risks) of Cyber War," *The New Yorker*, June 6, 2012. http://www.newyorker.com/news/daily-comment/the-rewards-and-risks-of-cyber-war.

Collins, Keith. "Net Neutrality Has Officially Been Repealed. Here's How That Could Affect You." *The New York Times*, June 11, 2018. https://www.nytimes.com/2018/06/11/technology/net-neutrality-repeal.html.

Conley, H.A., J. Mina, R. Stefanov, M. Vladimirov, *The Kremlin Playbook: Understanding Russian Influence in Central and Eastern Europe*. Boulder CO: Rowman & Littlefield, 2016.

Connell, Michael and Sarah Vogler. "Russia's Approach to Cyber Warfare." *CNA Analysis and Solutions*, March 2017. https://www.cna.org/CNA_files/PDF/DOP-2016-U-014231-1Rev.pdf.

Constantin, Lucian. "New Havex Malware Variants Target Industrial Control System and SCADA Users." *PC World*, June 24, 2014. http://www.pcworld.com/article/2367240/new-havex-malware-variants-target-industrial-control-system-and-scada-users.html and Havex Trojan: ICS-ALERT-14-176-02A.

Cook, Col. James. "'Cyberation' and Just War Doctrine: A Response to Randall Dipert." *Journal of Military Ethics*, Volume 9, Issue 4, 2010. https://www.law.upenn.edu/live/files/1701-cook-cyberation.

Crail, Peter. "IAEA: Syria Tried to Build Nuclear Reactor." *Arms Control Association*, March 2009. https://www.armscontrol.org/act/2009-03/iaea-syrian-reactor-explanation-suspect.

Czekaj, Matthew. "Russia's Hybrid War Against Poland." *Eurasia Daily Monitor*, Volume 12, Issue 80, April 29, 2015. https://jamestown.org/program/russias-hybrid-war-against-poland/.

Davidson, Jacob. "China Accuses U.S. of Hypocrisy on Cyberattacks." *Time*, July 1, 2013. http://world.time.com/2013/07/01/china-accuses-u-s-of-hypocrisy-on-cyberattacks/.

Davis, Dai. "Hacktivism: Good or Evil?" *Computer Weekly*, March 14, 2014. http://www.computerweekly.com/opinion/Hacktivism-Good-or-Evil.

Davis, Elizabeth Van Wie. "Can Technology Help China Rebuild Social Trust?" *Fair Observer*, August 2019. https://www.fairobserver.com/region/asia_pacific/china-social-credit-system-surveillance-technology-asia-pacific-news-32411/.

Davis, Elizabeth Van Wie. "China's Cyberwarfare Finds New Targets." *Fair Observer*, October 27, 2017. https://www.fairobserver.com/region/asia_pacific/china-cyberwarfare-cybersecurity-asia-pacific-news-analysis-04253/

Davis, Elizabeth Van Wie. "Don't Underestimate North Korea's Cyber Efforts." Fair Observer, March 21, 2018. https://www.fairobserver.com/region/asia_pacific/north-korea-cyberattacks-cybersecurity-asia-pacific-news-analysis-15400/.

Decree of the President of the Russian Federation. "On approval of the Doctrine of Information Security of the Russian Federation." December 5, 2016. http://publication.pravo.gov.ru/Document/View/0001201612060002.

Denning, Dorothy. *Information Warfare and Security*. New York: Addison-Wesley Professional, 1998.

Dent, Steve. "Report: Russia Hacked Election Systems in 39 US States." *Engadget*, June 13, 2017. https://www.engadget.com/2017/06/13/report-russia-hacked-election-systems-in-39-us-states/.

DeTrani, Joseph R. "Cyberspace: A Global Threat to Peace." *Asia Times*, October 28, 2013.

Deutsche Welle. "Merkel Testifies on NSA Spying Affair." *Deutsche Welle*, February 16, 2017. http://www.dw.com/en/merkel-testifies-on-nsa-spying-affair/a-37576690.

DeWeese, Steve, Bryan Krekel, George Bakos, and Christopher Barnett. *US-China Economic and Security Review Commission Report on the Capability of the People's Republic of China to Conduct Cyber Warfare and Computer Network Exploitation*. West Falls Church, VA: Northrop Grumman Corporation Information Systems Sector, 2009.

DeYoung, Karen, Ellen Nakashima, and Emily Rauhala. "US-Trump Signed Presidential Directive Ordering Actions to Pressure North Korea." *The Washington Post*, September 30, 2017. https://www.washingtonpost.com/world/national-security/trump-signed-presidential-directive-ordering-actions-to-pressure-north-korea/2017/09/30/97c6722a-a620-11e7-b14f-f41773cd5a14_story.html?utm_term=.5d67f8804aa6.

Dilanian, Ken. "Watch Out. North Korea Keeps Getting Better at Hacking." *ABC News*, February 20, 2018. https://www.nbcnews.com/news/north-korea/watch-out-north-korea-keeps-getting-better-hacking-n849381.

Dipert, Randall R. "The Ethics of Cyberwarfare." *Journal of Military Ethics*, Volume 9, Issue 4, 2010. http://dx.doi.org/10.1080/15027570.2010.536404.

Downes, Larry. "On Internet Regulation, The FCC Goes Back To The Future." *Forbes*, March 12, 2018. https://www.forbes.com/sites/larrydownes/2018/03/12/the-fcc-goes-back-to-the-future/#56d1e3d05b2e.

Drogin, Bob. "Russians Seem to Be Hacking into Pentagon / Sensitive Information Taken—But Nothing Top Secret." *Los Angeles Times*, October 7, 1999.

Dugin, Alexander (2012). The Fourth Political Theory. Translated by Mark Sleboda; Michael Millerman. Arktos Media.

Dugin, Alexander. *The Fourth Political Theory*. New York: Arktos Media, 2010.

Fischer, Sabine. "The Donbas Conflict: Opposing Interests and Narratives, Difficult Peace Process." *SWP Research Paper* 2019/RP 05, April 2019. https://www.swp-berlin.org/10.18449/2019RP05/.

Franke, Don. *Cyber Security Basics: Protect Your Organization by Applying the Fundamentals*. Scotts Valley, CA: CreateSpace Independent Publishing Platform, 2016.

Freedom House. "Freedom on the Net 2015." *Freedom House*, October 2015. https://freedomhouse.org/sites/default/files/FOTN%202015%20Full%20Report.pdf.

Fruhlinger, Josh. "Top Cybersecurity Facts, Figures and Statistics for 2020." *CSO Cyber Security Report*, March 9, 2020. https://www.csoonline.com/article/3153707/security/top-5-cybersecurity-facts-figures-and-statistics.html.

Gabowitsch, Mischa. *Protest in Putin's Russia*. Cambridge: Polity Press, 2017.

Galeotti, Mark. "The Kremlin's Newest Hybrid Warfare Asset." *Foreign Policy*, June 12, 2017. https://foreignpolicy.com/2017/06/12/how-the-world-of-spies-became-a-gangsters-paradise-russia-cyberattack-hack/

Galeotti, Mark. "Crimintern: How the Kremlin Uses Russia's Criminal Networks in Europe." *European Council on Foreign Relations*, April 18, 2017. https://www.ecfr.eu/publications/summary/crimintern_how_the_kremlin_uses_russias_criminal_networks_in_europe.

Gallagher, Sean. "Researchers Claim China Trying to Hack South Korean Missile Defense Efforts." *ARS Technica*, April 21, 2017. https://arstechnica.com/security/2017/04/researchers-claim-china-trying-to-hack-south-korea-missile-defense-efforts/.

Gan, Nectar. "What Do We Actually Know about China's Mysterious Spy Agency?" *South China Morning Post*, December 22, 2018. https://www.scmp.com/news/china/politics/article/2179179/what-do-we-actually-know-about-chinas-mysterious-spy-agency.

Gellman, Barton and Ellen Nakashima. "U.S. Spy Agencies Mounted 231 Offensive Cyber-Operations in 2011, documents show." *The Washington Post*, August 30, 2013. https://www.washingtonpost.com/world/national-security/us-spy-agencies-mounted-231-offensive-cyber-operations-in-2011-documents-show/2013/08/30/d090a6ae-119e-11e3-b4cb-fd7ce041d814_story.html?utm_term=.e963d1e06ce4.

Gellman, Barton, Ashkan Soltani, and Andrea Peterson. "How We Know the NSA Had Access to Internal Google and Yahoo Cloud Data." *The Washington Post*, November 4, 2013. http://www.washingtonpost.com/blogs/the-switch/wp/2013/11/04/how-we-know-the-nsa-had-access-to-internal-google-and-yahoo-cloud-data/?tid=hpModule_88854bf0-8691-11e2-9d71-f0feafdd1394&hpid=z11.

Gerden, Eugene. "$500 Million for New Russian Cyber Army." *SC Magazine*, November 6, 2014. http://www.scmagazineuk.com/500-million-for-new-russian-cyberarmy/article/381720/

Gilbert, David. "Cyber War—Just the Beginning of a New Military Era." *International Business Times*, April 26, 2013. http://www.ibtimes.co.uk/articles/461688/20130426/cyber-war-beginning-new-military-era.htm.

Girard, Rene. "On War and Apocalypse." *First Things*, August 2009. https://www.firstthings.com/article/2009/08/on-war-and-apocalypse.

Girard, René. *Battling to the End: Conversations With Benoît Chantre*. East Lansing, MI: Michigan State University Press, 2009.

Gjelten, Tom. "Seeing the Internet as an 'Information Weapon'." *National Public Radio*, September 23, 2010. http://www.npr.org/templates/story/story.php?storyId=130052701. Retrieved September 23, 2010.

Gjelten, Tom. "First Strike: US Cyber Warriors Seize the Offensive." *World Affairs*, January 23, 2013. http://www.worldaffairsjournal.org/article/first-strike-us-cyber-warriors-seize-offensive

Glanz, James and John Markoff, "Vast Hacking by a China Fearful of the Web." *The New York Times*, December 4, 2010. http://www.nytimes.com/2010/12/05/world/asia/05wikileaks-china.html?pagewanted=all&_r=1&.

Gold, Josh. "Two Incompatible Approaches to Governing Cyberspace Hinder Global Consensus." *Leiden Security and Global Affairs*, May 16, 2019. https://www.leidensafetyandsecurityblog.nl/articles/two-incompatible-approaches-to-governing-cyberspace-hinder-global-consensus.

Goldsmith, Jack. "The Precise (and Narrow) Limits On U.S. Economic Espionage." *Lawfare*, March 23, 2015. https://www.lawfareblog.com/precise-and-narrow-limits-us-economic-espionage

Gorman, Siobhan. "Electricity Grid in U.S. Penetrated By Spies." *The Wall Street Journal*, April 8, 2009. http://online.wsj.com/article/SB123914805204099085.html.

Gorshenin, Vadim. "Russia to Create Cyber-Warfare Units." *Pravda*, August 28, 2013. http://english.pravda.ru/russia/politics/29-08-2013/125531-cyber_warfare-0/.

Goud, Naveen. "China Cyberattacks Indian SUKHOI 30 Jet Fighters!" *Cybersecurity Insiders*, June 5, 2017. https://www.cybersecurity-insiders.com/china-cyber-attacks-indian-sukhoi-30-jet-fighters/.

Graff, Garrett M. "Indicting 12 Russian Hackers Could Be Mueller's Biggest Move Yet." *Wired*, July 13, 2018. https://www.wired.com/story/mueller-indictment-dnc-hack-russia-fancy-bear/.

Graham, Bradley. "Hackers Attack Via Chinese Web Sites." *The Washington Post*, August 25, 2005. http://www.washingtonpost.com/wp-dyn/content/article/2005/08/24/AR2005082402318.html.

Gray, C. *Another Bloody Century—Future Warfare* London: Weidenfeld/Nicolson, 2005.

Greenberg, Andy. "How an Entire Nation Became Russia's Test Lab For Cyberwar." *Wired*, June 20, 2017. https://www.kyivpost.com/ukraine-politics/wired-entire-nation-became-russias-test-lab-cyberwar.html.

Greenberg, Andy. "New Group of Iranian Hackers Linked to Destructive Malware." *Wired*, September 20, 2017. https://www.wired.com/story/iran-hackers-apt33/.

Greenwald, Glenn and Ewen MacAskill. "Obama Orders US to Draw Up Overseas Target List for Cyber-Attacks." *The Guardian*, June 7, 2013. http://www.theguardian.com/world/2013/jun/07/obama-china-targets-cyber-overseas

Harknett, Richard J. "United States Cyber Command's New Vision: What It Entails and Why It Matters." *Lawfare*, March 23, 2018. https://www.lawfareblog.com/united-states-cyber-commands-new-vision-what-it-entails-and-why-it-matters.

Hayden, Michael V. *The Assault on Intelligence: American National Security in an Age of Lies*. London: Penguin Press, 2018.

Heickerö, Roland. *Emerging Cyber Threats and Russian Views on Information Warfare and Information Operations*. Stockholm: Swedish Defence Research Agency, 2010.

Hersh, Seymour M. "The Online Threat: Should We Be Worried about a Cyber War?" *The New Yorker*, November 1, 2010. http://www.newyorker.com/reporting/2010/11/01/101101fa_fact_hersh#ixzz1LP462Ulr.

Higgins, Andrew. "Effort to Expose Russia's 'Troll Army' Draws Vicious Retaliation." *The New York Times*, May 31, 2016. http://www.nytimes.com/2016/05/31/world/europe/russia-finland-nato-trolls.html?_r=0.

Hille, Katherin and Christian Shepard. "Taiwan: Concern Grows over China's Invasion Threat." *Financial Times*, January 8, 2020. https://www.ft.com/content/e3462762-3080-11ea-9703-eea0cae3f0de.

Horsley, Jamie. "China's Orwellian Social Credit Score Isn't Real: Blacklists and Monitoring Systems are Nowhere Close to Black Mirror Fantasies." *Foreign Policy*, November 16, 2018. https://foreignpolicy.com/2018/11/16/chinas-orwellian-social-credit-score-isnt-real/.

Ignatius, David. "Russia is Pushing to Control Cyberspace. We Should all be Worried." *The Washington Post*, October 24, 2017. https://www.washingtonpost.com/opinions/global-opinions/russia-is-pushing-to-control-cyberspace-we-should-all-be-worried/2017/10/24/7014bcc6-b8f1-11e7-be94-fabb0f1e9ffb_story.html.

Joffee, Julia. "How State-Sponsored Blackmail Works in Russia." *The Atlantic*, January 11, 2017. https://www.theatlantic.com/international/archive/2017/01/kompromat-trump-dossier/512891/.

Joselow, Gabe. "Election Cyberattacks: Pro-Russia Hackers Have Been Accused in Past." *NBC News*, November 3, 2016. https://www.nbcnews.com/mach/technology/election-cyberattacks-pro-russia-hackers-have-been-accused-past-n673246.

Jun, Jenny, Scott LaFoy, and Ethan Sohn. "North Korea's Cyber Operations: Strategy and Responses." The Center for Strategic and International Studies, December 2015.

Kania, Elsa B. "Made in China 2025, Explained." *The Diplomat*, February 1, 2019. https://thediplomat.com/2019/02/made-in-china-2025-explained/.

Kaplan, Frank. *Dark Territory: The Secret History of Cyber War*. New York: Simon & Schuster, 2017.

Karber, Phillip and Joshua Thibeault. "Russia's New-Generation Warfare." *Association of the United States Army*, May 20, 2016. https://www.ausa.org/articles/russia%E2%80%99s-new-generation-warfare.

Kaufmann, William. "The Evolution of Deterrence, 1945–1958." unpublished RAND research, 1958.

Kelleher, Kevin. "Microsoft Says Russia has Already Tried to Hack 3 Campaigns in the 2018 Election." *Fortune*, July 19, 2018. http://fortune.com/2018/07/19/microsoft-russia-hack-2018-election-campaigns/.

Kennan, George F. *Policy Planning Staff Memorandum 269*. Washington, DC: U.S. State Department, May 4, 1948. http://academic.brooklyn.cuny.edu/history/johnson/65ciafounding3.htm.

Knake, Robert K. "A Cyberattack on the US Power Grid," Contingency Planning Memorandum No. 31, *Council on Foreign Relations*, April 3, 2017. https://www.cfr.org/report/cyberattack-us-power-grid.

Kobie, Nicole. "The Complicated Truth about China's Social Credit System." *Wired*, January 21, 2019. https://www.wired.co.uk/article/china-social-credit-system-explained.

Koerner, Brendan I. "Inside the Cyberattack That Shocked the US Government." *Wired*, October 13, 2016. https://www.wired.com/2016/10/inside-cyberattack-shocked-us-government/.

Kosaka, Tetsuro. "China's Military Reorganization could be a Force for Destabilization." *Nikkei Asian Review*, January 28, 2016. https://asia.nikkei.com/Politics/China-s-military-reorganization-could-be-a-force-for-destabilization.

Kostka, Genia. "What do People in China Think About 'Social Credit' Monitoring?" *The Washington Post*, March 21, 2019. https://www.washingtonpost.com/politics/2019/03/21/what-do-people-china-think-about-social-credit-monitoring/?utm_term=.49fe491dd67b.

Kowalewski, Annie. "China's Evolving Cybersecurity Strategy." *Georgetown Security Studies Review*, October 27, 2017. http://georgetownsecuritystudiesreview.org/2017/10/27/chinas-evolving-cybersecurity-strategy/.

Kramer, Andrew. "Russian General Pitches 'Information' Operations as a Form of War." *The New York Times*, March 2, 2019. https://www.nytimes.com/2019/03/02/world/europe/russia-hybrid-war-gerasimov.html.

Kramer, Franklin D., Stuart H. Starr, and Larry Wentz, ed. *Cyberpower and National Security* Washington, DC: National Defense University & Potomac Books Inc., 2009.

Kravchenko, Stepan. "Russia More Prey Than Predator to Cyber Firm Wary of China." *Bloomberg News*, August 25, 2016. https://www.bloomberg.com/news/articles/2016-08-25/russia-more-prey-than-predator-to-cyber-firm-wary-of-china.

Ku, Julian. "Forcing China to Accept that International Law Restricts Cyber Warfare May Not Actually Benefit the U.S." *Lawfare*, August 25, 2017. https://www.lawfareblog.com/forcing-china-accept-international-law-restricts-cyber-warfare-may-not-actually-benefit-us.

Lee, Dave. "'Red October' Cyber-Attack Found by Russian Researchers." *BBC News*, January 14, 2013. http://www.bbc.com/news/technology-21013087.

Lewis, James A. and Katrina Timlin. *Cybersecurity and Cyberwarfare: Preliminary Assessment of National Doctrine and Organization*. Washington, DC: Center for Strategic and International Studies, 2011.

Libicki, Martin. *Cyberdeterrence and Cyberwar*. Santa Monica, CA: Rand Publishing, 2009.

Lizza, Ryan. "State of Deception—Why Won't the President Rein in the Intelligence Community?" *The New Yorker*, December 16, 2013. http://www.newyorker.com/reporting/2013/12/16/131216fa_fact_lizza?currentPage=all.

Luo, Yan. Zhijing Yu, and Nicholas Shepherd. "China's Ministry of Public Security Issues New Personal Information Protection Guideline." *Inside Privacy*, April 19, 2019. https://www.insideprivacy.com/data-security/chinas-ministry-of-public-security-issues-new-personal-information-protection-guideline/.

Mackinnon, Amy. "Tinder and the Russian Intelligence Services: It's a Match! Will Facebook and Twitter be Next?" *Foreign Policy*, June 7, 2019. https://foreignpolicy.com/2019/06/07/tinder-and-the-russian-intelligence-services-its-a-match/.

Mandiant, *APT1: Exposing One of China's Cyber Espionage Units Mandiant Report.* Milpitas, CA: FireEye, 2013. https://www.fireeye.com/content/dam/fireeye-www/services/pdfs/mandiant-apt1-report.pdf.

Maréchal, Nathalie. "Are You Upset About Russia Interfering With Elections?" *Slate*, March 20, 2017. http://www.slate.com/articles/technology/future_tense/2017/03/russia_s_election_interfering_can_t_be_separated_from_its_domestic_surveillance.html.

Markoff, John. "Before the Gunfire, Cyberattacks." *The New York Times*, August 12, 2008. http://www.nytimes.com/2008/08/13/technology/13cyber.html?em&_r=0

Markoff, John and Andrew E. Kramer. "U.S. and Russia Differ on a Treaty for Cyberspace." *The New York Times*, June 27, 2009, http://www.nytimes.com/2009/06/28/world/28cyber.html.

Masters, Jonathan. "Confronting the Cyber Threat." *Council on Foreign Relations Publication*, March 17, 2011.

Matsakis, Louise. "What Happens if Russia Cuts Itself Off From the Internet?" *Wired*, February 12, 2019. https://www.wired.com/story/russia-internet-disconnect-what-happens/.

Mattis, Lieutenant General James N. USMC, and Lieutenant Colonel Frank Hoffman, USMCR (Ret.). "Future Warfare: The Rise of Hybrid Wars." *Proceedings Magazine*, Volume 132, November 2005. http://milnewstbay.pbworks.com/f/MattisFourBlockWarUSNINov2005.pdf.

Mazzetti, Mark and Katie Benner. "12 Russian Agents Indicted in Mueller Investigation." *The New York Times*, July 13, 2018. https://www.nytimes.com/2018/07/13/us/politics/mueller-indictment-russian-intelligence-hacking.html.

Meeuwisse, Raef. *Cybersecurity for Beginners.* London: Cyber Simplicity Ltd, 2017.

Mehrotra, Kartikay. "Jailed Russian of Interest in U.S. Election Probe, Official Says." *Bloomberg News*, August 24, 2018. https://www.bloomberg.com/news/articles/2018-08-24/quiet-jailed-russian-said-of-interest-in-u-s-election-meddling.

Meister, Stefan. "The 'Lisa Case': Germany as a Target of Russian Disinformation." *NATO Review*, July 25, 2016. https://www.nato.int/docu/review/articles/2016/07/25/the-lisa-case-germany-as-a-target-of-russian-disinformation/index.html.

Menn, Joseph. "China-Based Campaign Breached Satellite, Defense Companies: Symantec." *Reuters*, June 19, 2018. https://www.reuters.com/article/us-china-usa-cyber/china-based-campaign-breached-satellite-defense-companies-symantec-idUSKBN1JF2X0.

Meyer, Paul. "Diplomatic Alternatives to Cyber-Warfare: A Near-Term Agenda." *The RUSI Journal*, Volume 157, Issue 1, p. 14–19, 2012.

Mikheeva, Katerina. "Why the Russian Ecommerce Market is Worth the Hassle for Western Companies." *Digital Commerce*, April 9, 2019. https://www.digitalcommerce360.com/2019/04/09/why-the-russian-ecommerce-market-is-worth-the-hassle-for-western-companies/.

Miller, Greg, Ellen Nakashima, and Adam Entous. "Obama's Secret Struggle to Punish Russia for Putin's Election Assault." *The Washington Post*, June 23, 2017. https://www.washingtonpost.com/graphics/2017/world/national-security/obama-putin-election-hacking/?tid=ss_mail&utm_term=.383942a7e764.

Mistreanu, Simina. "Life Inside China's Social Credit Laboratory." *Foreign Policy*, April 3, 2018. https://foreignpolicy.com/2018/04/03/life-inside-chinas-social-credit-laboratory/

Monaghan, Andrew. "Putin's Way of War: The 'War' in Russia's 'Hybrid Warfare'." *Parameters*, Winter 2015–16. https://ssi.armywarcollege.edu/pubs/parameters/issues/winter_2015-16/9_monaghan.pdf.

Morgan, Steve. "Top 5 Cybersecurity Facts, Figures and Statistics for 2018." *CSO Cyber Security Report*, January 23, 2018. https://www.csoonline.com/article/3153707/security/top-5-cybersecurity-facts-figures-and-statistics.html.

Morse, Andrew. "Snowden: Alleged NSA Attack is Russian Warning." *Cnet*, August 16, 2016. https://www.cnet.com/news/snowden-nsa-hack-russia-warning-election-democratic-party/.

Mozur, Paul. "One Month, 500,000 Face Scans: How China Is Using A.I. to Profile a Minority." *The New York Times*, May 5, 2019. https://www.nytimes.com/2019/04/14/technology/china-surveillance-artificial-intelligence-racial-profiling.html?action=click&module=RelatedLinks&pgtype=Article.

Mueller, III, Robert S., Rosalind S Helderman, Matt Zapotosky, and US Department of Justice. Special Counsel's Office, *Report on the Investigation into Russian Interference in the 2016 Presidential Election*. New York: Scribner, 2019.

Mulvenon, James C. and Gregory J. Rattray. eds. *Addressing Cyber Instability*. Morrisville, NC: Lulu Press, 2012.

Nakamura, Kennon H. and Matthew C Weed. *US Public Diplomacy: Background and Current Issues*. Washington, DC: Congressional Research Service December 18, 2009.

Nakashima, Ellen. "Military Leaders Seek More Clout for Pentagon's Cyber Command Unit." *The Washington Post*, May 1, 2012. https://www.washingtonpost.com/world/national-security/military-officials-push-to-elevate-cyber-unit-to-full-combatant-command-status/2012/05/01/gIQAUud1uT_story.html?utm_term=.82a37e29370b.

Nakashima, Ellen. "Newly Identified Computer Virus, Used for Spying, is 20 Times Size of Stuxnet." *The Washington Post*, May 28, 2012. http://www.washingtonpost.com/world/national-security/newly-identified-computer-virus-used-for-spying-is-20-times-size-of-stuxnet/2012/05/28/gJQAWa3VxU_story.html.

Nakashima, Ellen. "Legal Memos Released on Bush-Era Justification for Warrantless Wiretapping." *The Washington Post*, September 6, 2014. https://www.was

hingtonpost.com/world/national-security/legal-memos-released-on-bush-era-justification-for-warrantless-wiretapping/2014/09/05/91b86c52-356d-11e4-9e92-0899b306bbea_story.html.

Nakashima, Ellen. "Russia Has Developed a Cyberweapon that can Disrupt Power Grids, According to New Research" *The Washington Post*, June 12, 2017. https://www.washingtonpost.com/world/national-security/russia-has-developed-a-cyber-weapon-that-can-disrupt-power-grids-according-to-new-research/2017/06/11/b91b773e-4eed-11e7-91eb-9611861a988f_s.tory.html?tid=ss_mail&utm_term=.35bafd178f13.

Nakashima, Ellen. "With Plan X, Pentagon Seeks to Spread U.S. Military Might to Cyberspace." *The Washington Post*, May 30, 2012. http://www.washingtonpost.com/world/national-security/with-plan-x-pentagon-seeks-to-spread-us-military-might-to-cyberspace/2012/05/30/gJQAEca71U_story.html.

Nakashima, Ellen, Greg Miller, and Julie Tate. "U.S., Israel Developed Flame Computer Virus to Slow Iranian Nuclear Efforts, Officials Say." *The Washington Post*, June 19, 2012. http://www.washingtonpost.com/world/national-security/us-israel-developed-computer-virus-to-slow-iranian-nuclear-efforts-officials-say/2012/06/19/gJQA6xBPoV_story.html.

NATO. "Lisbon Summit Declaration: Issued by the Heads of State and Government Participating in the Meeting of the North Atlantic Council in Lisbon." *North Atlantic Treaty Organization*, November 20, 2010. features/2010/0410_cybersec/docs/CYberFactSheet%20UPD.

Naymushin, Ilya. "Russian Dam Disaster Kills 10, Scores Missing." *Reuters*, August 16, 2009. https://www.reuters.com/article/us-russia-accident-sb/russian-dam-disaster-kills-10-scores-missing-idUSTRE57G0M120090817.

Nechepurenko, Ivan and Michael Schwirtz. "What We Know About Russians Sanctioned by the United States." *The New York Times*, February 17, 2018. https://www.nytimes.com/2018/02/17/world/europe/russians-indicted-mueller.html.

Negroponte, John D. et al., "Defending an Open, Global, Secure, and Resilient Internet," *The Council on Foreign Relations*, June 2013. https://www.cfr.org/content/publications/attachments/TFR70_cyber_policy.pdf.

Neudert, Lisa-Maria N. "Computational Propaganda in Germany: A Cautionary Tale." In *Project on Computational Propaganda*, edited by Samuel Woolley and Philip N. Howard. Oxford: Oxford University, 2017. http://comprop.oii.ox.ac.uk/wp-content/uploads/sites/89/2017/06/Comprop-Germany.pdf.

Novosti, R.I.A. and Mikhail Fomichev. "Russia to Press for International Internet Behavior Code to Fight Emerging Threats." *Russia Times*, August 01, 2013. http://rt.com/politics/russia-internet-international-code-893/.

Nye, Jr., Joseph S. *The Future of Power in the 21st Century*. New York: Public Affairs Press, 2011.

Oliphant, Roland, Rory Mulholland, Justin Huggler, and Senay Boztas. "How Vladimir Putin and Russia are Using Cyber Attacks and Fake News to Try to Rig Three Major European Elections this Year." *The Telegraph*, February 11, 2017. http://www.telegraph.co.uk/news/2017/02/13/vladimir-putin-russia-using-cyber-attacks-fake-news-try-rig/.

Osnos, Evan, David Remnick, and Joshua Yaffa. "Trump, Putin and the New Cold War." *New Yorker*, March 6, 2017. http://www.newyorker.com/magazine/2017/03/06/trump-putin-and-the-new-cold-war.

Owens, William A., Kenneth W. Dam, and Herbert S. Lin. eds., *Technology, Policy, Law, and Ethics Regarding US Acquisition and Use of Cyberattack Capabilities*. Washington, DC: National Academies Press, 2009.

Peng Guangqiang and Yao Youzhi. eds., *The Science of Military Strategy*. Beijing: Military Science Publishing House, English edition, 2005.

Perlroth, Nicole, Jeff Larson, and Scott Shane. "Revealed: The NSA's Secret Campaign to Crack, Undermine Internet Security." *The New York Times*, September 5, 2013. http://www.propublica.org/article/the-nsas-secret-campaign-to-crack-undermine-internet-encryption.

Petraeus, David. "Cyber Changed War, But The Causes And Conduct Of Conflict Remain Human" *The World Post*, March 29, 2017. https://www.belfercenter.org/publication/cyber-changed-war-causes-and-conduct-conflict-remain-human.

Prakash, Abishur. "Facial Recognition Cameras and AI: 5 Countries With the Fastest Adoption." *Robotics Business Review*, December 21, 2018. https://www.roboticsbusinessreview.com/ai/facial-recognitio.n-cameras-5-countries/.

Putin, Vladimir. *Russia and the Changing World*. February 27, 2012. https://www.rt.com/politics/official-word/putin-russia-changing-world-263/.

Quinn, Tyler. "The Bear's Side of the Story: Russian Political and Information Warfare." *RealClearDefense*, June 27, 2018. https://www.realcleardefense.com/articles/2018/06/27/the_bears_side_of_the_story_russian_political_and_information_warfare_113564.html.

Rattray, Gregory J., Chris Evans, and Jason Healey. "American Security in the Cyber Commons." In *Contested Commons: The Future of American Power in a Multipolar World*. edited by Abraham M. Denmark and Dr. James Mulvenon. Washington, DC: Center for a New American Security, January 2010.

Raud, Mikk. "China and Cyber: Attitudes, Strategies, Organisation." *NATO Cooperative Cyber Defence Centre of Excellence*, 2016. https://ccdcoe.org/multimedia/national-cyber-security-organisation-china.html.

Rauscher, Karl and Andrey Korotkov. *Working Towards Rules for Governing Cyber Conflict: Rendering the Geneva and Hague Conventions in Cyberspace*. Honolulu HI: East West Center, February 3, 2011.

Reuters. "Russian Accused of Hacking Extradited to U.S. from Spain." *Reuters*, February 2, 2018. https://www.reuters.com/article/us-usa-cyber-levashov/russian-accused-of-hacking-extradited-to-u-s-from-spain-idUSKBN1FM2RG.

Reuters. "Russian Hackers Accused of Targeting UN Chemical Weapons Watchdog, MH17 Files." *Australian Broadcasting Corporation*, October 4, 2018. https://www.abc.net.au/news/2018-10-04/russia-tried-to-hack-un-chemical-weapons-watchdog-netherlands/10339920.

Reynolds, Joe. "China's Evolving Perspectives on Network Warfare: Lessons from the Science of Military Strategy." *China Brief*, Volume 15, Issue 8, April 16, 2015 https://jamestown.org/program/chinas-evolving-perspectives-on-network-warfare-lessons-from-the-science-of-military-strategy/#.V1BM2_krK70.

Rid, Thomas. "Cyber War and Peace: Hacking Can Reduce Real World Violence." *Foreign Affairs*, Volume 92, Issue 6, p. 77–87, November/December 2013.

Riley, Michael and Jordan Robertson. "Russian Cyber Hacks on U.S. Electoral System Far Wider Than Previously Known." *Bloomberg News*, June 13, 2017. https://www.bloomberg.com/news/articles/2017-06-13/russian-breach-of-39-states-threatens-future-u-s-elections.

Robbins, Gary. "Why are China and Russia Getting Hit Hard by Cyber Attack, But Not the U.S.?" *The San Diego Union-Tribune*, May 15, 2017. http://www.sandiegouniontribune.com/news/cyber-life/sd-me-ransomware-update-20170515-story.html.

Robinson, Paul. ed. *Just War in Comparative Perspective*. Abingdon-on-Thames, UK: Routledge, 2003.

Robles, Frances. "Russian Hackers Were 'In a Position' to Alter Florida Voter Rolls, Rubio Confirms." *The New York Times*, April 26, 2019. https://www.nytimes.com/2019/04/26/us/florida-russia-hacking-election.html.

Rowberry, Ariana. "Sixty Years of 'Atoms for Peace' and Iran's Nuclear Program." *The Brookings Institute*, December 18, 2013. https://www.brookings.edu/blog/up-front/2013/12/18/sixty-years-of-atoms-for-peace-and-irans-nuclear-program/.

Russian Federation Ministry of Foreign Affairs. "Convention on International Information Security," September 22, 2011. https://carnegieendowment.org/files/RUSSIAN-DRAFT-CONVENTION-ON-INTERNATIONAL-INFORMATION-SECURITY.pdf.

Russian Federation Office of the President, *The Military Doctrine of the Russian Federation*. No. Pr.-2976, December 25, 2014. https://rusemb.org.uk/press/2029.

Sadyki, Marina. "National Report on E-commerce Development in Russia." *UN Industrial Development Organization*, 2017. https://www.unido.org/api/opentext/documents/download/9920890/unido-file-9920890

Sanger, David E. "Obama Order Sped Up Wave of Cyberattacks Against Iran." *The New York Times*, June 1, 2012. http://www.nytimes.com/2012/06/01/world/middleeast/obama-ordered-wave-of-cyberattacks-against-iran.html?pagewanted=all&_r=0.

Sanger, David E. "Chinese Curb Cyberattacks on U.S. Interests, Report Finds." *The New York Times*, June 20, 2016. https://www.nytimes.com/2016/06/21/us/politics/china-us-cyber-spying.html.

Sanger, David E. *The Perfect Weapon: War, Sabotage, and Fear in the Cyber Age*. New York: Crown Publishing Group, 2018.

Sanger, David E. "With Spy Charges, U.S. Draws a Line That Few Others Recognize." *The New York Times*, May 19, 2014. http://www.nytimes.com/2014/05/20/us/us-treads-fine-line-in-fighting-chinese-espionage.html?hp&_r=0.

Sanger, David E. and Eric Schmitt. "U.S. Cyberweapons, Used Against Iran and North Korea, Are a Disappointment Against ISIS." *The New York Times*, June 12, 2017. https://www.nytimes.com/2017/06/12/world/middleeast/isis-cyber.html?smid=nytcore-ipad-share&smprod=nytcore-ipad&_r=0.

Sanger, David E. and Nicole Perlroth. "U.S. Escalates Online Attacks on Russia's Power Grid." *The New York Times*, June 15, 2019. https://www.nytimes.com/2019/06/15/us/politics/trump-cyber-russia-grid.html.

Sanger, David E. and William J Broad. "Hand of U.S. Leaves North Korea's Missile Program Shaken." *The New York Times*, April 18, 2017. https://www.nytimes.com/2017/04/18/world/asia/north-korea-missile-program-sabotage.html?_r=0 .

Sanger, David E. and William J Broad. "Pentagon Suggests Countering Devastating Cyberattacks With Nuclear Arms." *The New York Times*, January 16, 2018. https://www.nytimes.com/2018/01/16/us/politics/pentagon-nuclear-review-cyberattack-trump.html.

Sanger, David E., and Nicole Perlroth. "N.S.A. Breached Chinese Servers Seen as Security Threat." *The New York Times*, March 22, 2014. https://www.nytimes.com/2014/03/23/world/asia/nsa-breached-chinese-servers-seen-as-spy-peril.html.

Sanger, David E., David D. Kirkpatrick, and Nicole Perlroth. "The World Once Laughed at North Korean Cyberpower. No More." *The New York Times*, October 15, 2017. https://www.nytimes.com/2017/10/15/world/asia/north-korea-hacking-cyber-sony.html.

Sanger, David E., Davis Barboza, and Nicole Perlroth. "Chinese Army Unit Is Seen as Tied to Hacking Against U.S." *The New York Times*, February 18, 2013. https://www.nytimes.com/2013/02/19/technology/chinas-army-is-seen-as-tied-to-hacking-against-us.html.

Saran, Cliff. "F-Secure Warns of Russian State-Supported Cyber Espionage: Russian State-Sponsored Hackers Work Office Hours and Target Western Governments, According to F-Secure Report." *Computer Weekly*, September 17, 2015. http://www.computerweekly.com/news/4500253704/F-Secure-warns-of-Russian-state-supported-cyber-espionage.

Schectman, Joel, Dustin Volz, and Jack Stubbs. "HP Enterprise Let Russia Scrutinize Cyberdefense System Used by Pentagon," *Reuters*, October 2, 2017. http://www.reuters.com/article/us-usa-cyber-russia-hpe-specialreport/special-report-hp-enterprise-let-russia-scrutinize-cyberdefense-system-used-by-pentagon-idUSKCN1C716M.

Schmitt, Michael N. "Grey Zones in the International Law of Cyberspace." *Yale Journal of International Law* Volume 42, Issue 2, 2017a.

Schmitt, Michael N. "Peacetime Cyber Responses and Wartime Cyber Operations under International Law: An Analytical Vade Mecum." *Harvard National Security Journal* Issue 8, p. 240–280, 2017b.

Schmitt, Michael N. ed. *Tallinn Manual on the International Law Applicable to Cyber Warfare*. New York: Cambridge University Press, 2013.

Seddon, Max and Henry Foy, "Russian Technology: Can the Kremlin Control the Internet? *Financial Times*, June 4, 2019. https://www.ft.com/content/93be9242-85e0-11e9-a028-86cea8523dc2.

Segal, Adam. " How China is Preparing for Cyberwar." *Christian Science Monitor*, March 20, 2017. http://www.csmonitor.com/World/Passcode/Passcode-Voices/2017/0320/How-China-is-preparing-for-cyberwar.

Shalal, Andrea. "Germany Challenges Russia Over Alleged Cyberattacks." *Reuters*, May 4, 2017. http://www.reuters.com/article/us-germany-security-cyber-russia-idUSKBN1801CA.

Shane, Scott, Mark Mazzetti, and Matthew Rosenberg. "WikiLeaks Releases Trove of Alleged C.I.A. Hacking Documents." *The New York Times*, March 7, 2017. https://www.nytimes.com/2017/03/07/world/europe/wikileaks-cia-hacking.html?_r=0.

Shane, Scott, Nicole Perlroth, and David E. Sanger. "Security Breach and Spilled Secrets Have Shaken the N.S.A. to Its Core." *The New York Times*, November 12, 2017. https://www.nytimes.com/2017/11/12/us/nsa-shadow-brokers.html.

Shanghai Cooperation Organization. *Agreement on Cooperation in Ensuring International Information Security between the Member States of the Shanghai Cooperation Organization*. June 16, 2009. file:///C:/Users/davis/Downloads/Agreement_on_Cooperation_in_Ensuring_International_Information_Security_between_the_Member_States_of_the_SCO.pdf.

Shanker, Thom. "Pentagon Is Updating Conflict Rules in Cyberspace." *The New York Times*, June 27, 2013. http://www.nytimes.com/2013/06/28/us/pentagon-is-updating-conflict-rules-in-cyberspace.html.

Sharikov, Pasha. "Cybersecurity in Russian-U.S. Relations." *Center for International and Security Studies at Maryland Policy Brief*, April 2013. https://spp.umd.edu/sites/default/files/2019-07/policy_brief_april_2013__sharikov.pdf.

Shear, Michael, Eric Schmitt, and Maggie Haberman. "Trump Approves Strike on Iran, but Then Abruptly Pulls Back." *The New York Times*, June 20, 2019. https://www.nytimes.com/2019/06/20/world/middleeast/iran-us-drone.html.

Shotter, James. "Czechs Fear Russian Fake News in Presidential Election." *Financial Times*, January 8, 2018. https://www.ft.com/content/c2b36cf0-e715-11e7-8b99-0191e45377ec.

Singer, P.W. and Allan Freedman. *Cybersecurity and Cyberwar: What Everyone Needs to Know*. New York: Oxford University Press, 2014.

Smith, David J. "Russian Cyber Capabilities, Policy and Practice." *inFocus Quarterly*, Winter 2014.

Smith, Rebecca. "U.S. Officials Push New Penalties for Hackers of Electrical Grid: Red Line Set for Cyberattacks on Infrastructure after Russian Agents Penetrated Utility Control Rooms." *The Wall Street Journal*, August 5, 2018. https://www.wsj.com/articles/u-s-officials-push-new-penalties-for-hackers-of-electrical-grid-1533492714.

Snowden, Edward. *Twitter*, August 16, 2016. https://twitter.com/snowden/status/765513776372342784?lang=en.

Snyder, Charley and Michael Sulmeyer. "The Department of Justice Makes the Next Move in the U.S.-Russia Espionage Drama." *Lawfare*, March 16, 2017. https://www.lawfareblog.com/department-justice-makes-next-move-us-russia-espionage-drama.

Sokol, Sam. "Russian Disinformation Distorted Reality in Ukraine. Americans Should Take Note." *Foreign Policy*, August 2, 2019. https://foreignpolicy.com/2019/08/02/russian-disinformation-distorted-reality-in-ukraine-americans-should-take-note-putin-mueller-elections-antisemitism/.

Soldatov, Andrei and Irina Borogan. *The Red Web: The Struggle Between Russia's Digital Dictators and the New Online Revolutionaries*. New York: Hatchette Book Group, 2017.

Soo, Zen. "Here's How China's New E-commerce Law will Affect Consumers, Platform Operators." *South China Morning Post*, January 1, 2019. https://www.scmp.com/tech/apps-social/article/2180194/heres-how-chinas-new-e-commerce-law-will-affect-consumers-platform.

Spencer, David. "Why the Risk of Chinese Cyberattacks Could Affect Everyone in Taiwan." *Taiwan News*, July 13, 2018. https://www.taiwannews.com.tw/en/news/3481423.

Sridharan, Vasudevan. "Russia Setting Up Cyber Warfare Unit Under Military." *International Business Times*, August 20, 2013. http://www.ibtimes.co.uk/articles/500220/20130820/russia-cyber-war-hack-moscow-military-snowden.htm.

Stelzenmüller, Constanze."The Impact of Russian Interference on Germany's 2017 Election." Testimony before the US Senate Select Committee on Intelligence, *Brookings Institute*, June 28, 2017. https://www.brookings.edu/testimonies/the-impact-of-russian-interference-on-germanys-2017-elections/.

Stiennon, Richard. *Surviving Cyberwar*. Lanham, MD: Government Institutes, 2010.

Stout, Kristie Lu. "Cyber Warfare: Who is China Hacking Now?" *CNN*, September 29, 2016. http://www.cnn.com/2016/09/29/asia/china-cyber-spies-hacking/index.html.

Stubbs, Jack, Joseph Menn, and Christopher Bing. "China Hacked Eight Major Computer Service Firms in Years-Long Attack." *Reuters*, June 26, 2019. https://www.reuters.com/article/us-china-cyber-cloudhopper-companies-exc/exclusive-china-hacked-eight-major-computer-services-firms-in-years-long-attack-idUSKCN1TR1D4.

Sukhankin, Sergey. "Russia's New Information Security Doctrine: Fencing Russia from the 'Outside World'?" *Eurasia Daily Monitor*, Volume 13, Issue 198, December 16, 2016. https://www.refworld.org/docid/5864c6b24.html.

Sulmeyer, Michael. "Cybersecurity in the 2017 National Security Strategy." *Lawfare*, December 19, 2017. https://www.lawfareblog.com/cybersecurity-2017-national-security-strategy.

Sun Tzu. *The Art of War*. London: Oxford University Press, 1963.

Sussman, Bruce. "Cyber Attack Motivations: Russia vs. China." *Secure World*, June 3, 2019. https://www.secureworldexpo.com/industry-news/why-russia-hacks-why-china-hacks.

Sutherland, Iain, Konstantinos Xynos, Andrew Jones, and Andrew Blyth. "The Geneva Conventions and Cyber-Warfare: A Technical Approach." *The RUSI Journal*, September 4, 2015.

Tait, Robert. "Czech Cyber-Attack: Russia Suspected of Hacking Diplomats' Emails." *The Guardian*, January 31, 2017. https://www.theguardian.com/world/2017/jan/31/czech-cyber-attack-russia-suspected-of-hacking-diplomats-emails.

Tait, Robert and Julian Borger. "Alleged Hacker held in Prague at Center of 'intense' US-Russia Tug of War." *The Guardian*, January 27, 2017. https://www.theguardian.com/technology/2017/jan/27/us-russia-hacking-yevgeniy-nikulin-linkedin-dropbox.

Thornton, Rod. "The Changing Nature of Modern Warfare: Responding to Russian Information Warfare." *The RUSI Journal*, Volume 160, Issue 4, 2015.

Tiezzi, Shannon. "Xi Jinping Leads China's New Internet Security Group." *The Diplomat*, February 28, 2014. https://thediplomat.com/2014/02/xi-jinping-leads-chinas-new-internet-security-group/.

Tikk, Eneken and Mika Kerttunen. "Parabasis: Cyber-diplomacy in Stalemate." *Norwegian Institute of International Affairs*, May 2018. https://nupi.brage.unit

.no/nupi-xmlui/bitstream/handle/11250/2569401/NUPI_Report_5_18_Tikk_Kerttunen.pdf?sequence=1&isAllowed=y.

Timberg, Craig. "Spreading Fake News Becomes Standard Practice for Governments across the World." *The Washington Post*, July 17, 2017. https://www.washingtonpost.com/news/the-switch/wp/2017/07/17/spreading-fake-news-becomes-standard-practice-for-governments-across-the-world/?hpid=hp_hp-cards_hp-card-technology%3Ahomepage%2Fcard&utm_term=.248d94c26b31.

Townsend, Kevin. "The United States and China—A Different Kind of Cyberwar." *Security Week*, January 7, 2019. https://www.securityweek.com/united-states-and-china-different-kind-cyberwar.

Trenin, Dmitri. "Russia and Germany: From Estranged Partners to Good Neighbors." *Carnegie Moscow Center*, June 2018. https://carnegieendowment.org/files/Article_Trenin_RG_2018_Eng.pdf.

UK House of Commons Digital, Culture, Media and Sport Committee. *Disinformation and 'fake news': Final Report Eighth Report of Session 2017–19 Report*. London: The House of Commons, February 14, 2019.

United Nations General Assembly. "International Code of Conduct for Information Security." Report of the United Nations General Assembly, 2015.

United Nations Office of Disarmament Affairs. "Developments in the Field of Information and Telecommunications in the Context of International Security." Report of the United Nations, A/54/213, December 2018.

United Nations Office of Drugs and Crime, "The Use of the Internet for Terrorist Purposes," Report of the United Nations, 2012.

US Cyber Command. *Achieve and Maintain Cyberspace Superiority: Command Vision for US Cyber Command*. March 23, 2018 https://assets.documentcloud.org/documents/4419681/Command-Vision-for-USCYBERCOM-23-Mar-18.pdf.

US Department of Defense. *US Cyber Command Fact Sheet*. Washington, DC, 2018.

US Department of Defense. "Evaluation of Possible Military Courses of Action," *Foreign Relations of the United States, 1961–63*, Volume X Cuba, January 1961–September 1962. https://history.state.gov/historicaldocuments/frus1961-63v10/d19.

US Department of Defense. *Strategy for Operating in Cyberspace*. Washington, DC, July 2011.

US Department of Defense. *The DOD Cyber Strategy*. Washington, DC, April 2015. https://www.defense.gov/Portals/1/features/2015/0415_ cyber-strategy/Final_2015_DoD_CYBER_STRATEGY_for_web.pdf.

US Department of Defense/Defense Science Board, *Task Force on Cyber Deterrence*, February 2017.

US Department of Homeland Security. *Homeland Security Presidential Directive 7: Critical Infrastructure Identification, Prioritization, and Protection*. Washington, DC, December 17, 2003.

US Department of Homeland Security. *Dam Sector Security Awareness Guide*. Washington, DC, 2007. https://www.dhs.gov/sites/default/files/publications/ip_dams_sector_securit_awareness_guide_508_0.pdf.

US Department of Homeland Security. *Distributed Denial of Service Defense (DDoSD)*. Washington, DC, November 21, 2016. https://www.dhs.gov/sites/defa

ult/files/publications/FactSheet%20DDoSD%20FINAL%20508%20OCC%20Cleared.pdf.

US Department of Homeland Security. *Alert (ICS-ALERT-17-206-01) CRASHOVERRIDE Malware*. Washington, DC, July 25, 2017. https://ics-cert.us-cert.gov/alerts/ICS-ALERT-17-206-01.

US Department of Justice. "Grand Jury Indicts Thirteen Russian Individuals and Three Russian Companies for Scheme to Interfere in the United States Political System." *Justice News*, February 16, 2018. https://www.justice.gov/opa/pr/grand-jury-indicts-thirteen-russian-individuals-and-three-russian-companies-scheme-interfere.

US Department of Justice. "US Charges Russian FSB Officers and Their Criminal Conspirators for Hacking Yahoo and Millions of Email Accounts." *DOJ Press Release*, March 15, 2017. https://www.justice.gov/opa/pr/us-charges-russian-fsb-officers-and-their-criminal-conspirators-hacking-yahoo-and-millions.

US Department of Justice. *US Indictment CRIMINAL NO. (18 U.S.C. §§ 2, 371, 1349, 1028A)*. Washington, DC, February 16, 2018. https://www.justice.gov/file/1035477/download.

US Department of the Interior, Office of the Inspector General. *US Bureau of Reclamation Selected Hydropower Dams at Increased Risk From Insider Threats*. Washington, DC, June 2018. https://www.doioig.gov/sites/doioig.gov/files/FinalEvaluation_ICSDams_Public.pdf.

US Deputy Secretary of Defense. *Memorandum, Subject: The Definition of Cyberspace*. Washington, DC, May 12, 2008.

US Director of National Intelligence. *Cyber Threat Intelligence Integration Center*. Washington, DC, February 2015. https://www.dni.gov/files/CTIIC/documents/CTIIC-Overview_for-unclass.pdf.

US Director of US Intelligence. *Assessing Russian Activities and Intentions in Recent US Elections*. Washington, DC, January 6, 2017.

US Joint Force Development, *Cyberspace Operations*. Washington, DC, June 8, 2018. https://www.jcs.mil/Portals/36/.ocuments/Doctrine/pubs/jp3_12.pdf?ver=2018-07-16-134954-150.

US National Counterintelligence Executive. *Foreign Spies Stealing US Economic Secrets in Cyberspace*. Washington, DC, October 2011. http://www.ncix.gov/publications/reports/fecie_all/Foreign_Economic_Collection_2011.pdf.

US Offices of Inspectors General of the Department of Defense, Department of Justice, Central Intelligence Agency, National Security Agency, and Office of the Director of National Intelligence. *Annex to the Report on the President's Surveillance Program*. Volume III, Washington, DC, July 10, 2009. https://oig.justice.gov/reports/2015/PSP-09-18-15-vol-III.pdf.

US President, Office of, *Cyber Space Policy Review: Assuring a Trusted and Resilient Information and Communications Infrastructure*. Washington, DC, 2009.

US President, Office of, *The Comprehensive National Cybersecurity Initiative*. Washington, DC, 2009.

US Presidential Executive Order, *Strengthening the Cybersecurity of Federal Networks and Critical Infrastructure*. Washington, DC, 2017.

US Presidential Policy Directive/PPD 20 *US Cyber Operations Policy*. 2012. https://fas.org/irp/offdocs/ppd/ppd-20.pdf.

US Senate, Committee on Intelligence. "Report of the Select Committee on Intelligence of the United States Senate on Russian Active Measures Campaigns and Interference in the 2016 US Election." Volume 4, Review of the Intelligence Community Assessment with Additional Views Washington, DC, April 2020. https://www.intelligence.senate.gov/sites/default/files/documents/Report_Volume4.pdf.

US White House Presidential Directive/NSC-63, *Critical Infrastructure Protection*. May 1998 https://www.documentcloud.org/documents/1513862-clinton-presidential-policy-directive.html

US White House, *National Security Strategy*, Washington, DC, May 2010.

US White House. *International Strategy for Cyberspace: Prosperity, Security, and Openness in a Networked World*. Washington, DC, May 2011. https://obamawhitehouse.archives.gov/sites/default/files/rss_viewer/international_strategy_for_cyberspace.pdf.

US White House, *Prosperity, Security, and Openness in a Networked World*. Washington, DC, May 11, 2012.

US White House, "US and Russia Sign Cyber Security Pact." *The Atlantic Council*, June 18, 2013. https://www.atlanticcouncil.org/blogs/natosource/us-and-russia-sign-cyber-security-pact.

US White House, *National Security Strategy* Washington, DC, 2017. https://www.whitehouse.gov/wp-content/uploads/2017/12/NSS-Final-12-18-2017-0905.pdf.

US White House, *Vulnerabilities Equities Policy and Process for the United States Government*. Washington, DC, November 15, 2017. https://www.whitehouse.gov/sites/whitehouse.gov/files/images/External%20-%20Unclassified%20VEP%20Charter%20FINAL.PDF.

Volz, Dustin. "Trump Administration Hasn't Briefed Congress on New Rules for Cyberattacks, Lawmakers Say: Some Lawmakers are Concerned they Lack Oversight of the Military's Increasing Use of Cyber Weapons." *The Wall Street Journal*, July 10, 2019. https://www.wsj.com/articles/trump-administration-hasnt-briefed-congress-on-new-rules-for-cyberattacks-lawmakers-say-11562787360.

von Clausewitz, Carl. *On War*. Original 1873, Reprinted Surrey, UK: Fab Press, 2010. https://www.clausewitz.com/readings/OnWar1873/BK1ch01.html.

Walsh, Declan and Ihsanullah Tipu Mehsud. "Civilian Deaths in Drone Strikes Cited in Report." *The New York Times*, October 22, 2013. http://www.nytimes.com/2013/10/22/world/asia/civilian-deaths-in-drone-strikes-cited-in-report.html?hpw&_r=0.

Wang Houqing and Zhang Xingye. eds. *The Science of Campaigns*. Beijing: National Defense University Press, May 2000.

Wang, Kevin Wei, and Jonathan Woetzel, et al. "Digital China: Powering the Economy to Global Competitiveness." *McKinsey Global Institute*, December 2017. https://www.mckinsey.com/featured-insights/china/digital-china-powering-the-economy-to-global-competitiveness.

Waterman, Shaun. "Analysis: Who Cyber Smacked Estonia?" *United Press International*, June 11, 2007. http://www.upi.com/Business_News/Security-Industry/2007/06/11/Analysis-Who-cyber-smacked-Estonia/UPI-26831181580439/.

Wentworth, Travis. "How Russia May Have Attacked Georgia's Internet." *Newsweek*, August 22, 2008. http://www.newsweek.com/how-russia-may-have-attacked-georgias-internet-88111.

Weston, Greg. "Foreign Hackers Attack Canadian Government: Computer Systems at 3 Key Departments Penetrated." *Canadian Broadcasting Corporation News*, February 16, 2011. http://www.cbc.ca/news/politics/foreign-hackers-attack-canadian-government-1.982618.

Whitlock, Craig. "Somali American Caught Up in a Shadowy Pentagon Counterpropaganda Campaign." *The Washington Post*, July 7, 2013. https://www.washingtonpost.com/world/national-security/somali-american-caught-up-in-a-shadowy-pentagon-counterpropaganda-campaign/2013/07/07/b3aca190-d2c5-11e2-bc43-c404c3269c73_story.html.

Wilber, Del Quentin. "Contractor Accused of Leaking NSA Document on Russian Hacking Pleads Guilty." *The Wall Street Journal*, June 26, 2018. https://www.wsj.com/articles/contractor-accused-of-leaking-nsa-document-on-russian-hacking-pleads-guilty-1530048276.

Windham, Robert. "Timeline: Ten Years of Russian Cyber Attacks on Other Nations." *NBC News*, December 18, 2016. http://www.nbcnews.com/news/us-news/timeline-ten-years-russian-cyber-attacks-other-nations-n697111.

Wintour, Patrick. "Russian Bid to Influence Brexit Vote Detailed in New US Senate Report." *The Guardian*, January 10, 2018. https://www.theguardian.com/world/2018/jan/10/russian-influence-brexit-vote-detailed-us-senate-report.

Wong, Sue-Lin and Michael Martina. "China Adopts Cyber Security Law in Face of Overseas Opposition." *Reuters Technology News*, November 7, 2016. http://www.reuters.com/article/us-china-parliament-cyber-idUSKBN132049.

Yadav, Yatish. "80,000 Cyberattacks on December 9 and 12 after Note Ban." *New India Express*, December 19, 2016. http://www.newindianexpress.com/nation/2016/dec/19/80000-cyber-attacks-on-december-9-and-12-after-note-ban-1550803.html.

You Ji. *China's Military Transformation*. Cambridge, UK: Polity Press, 2016.

Yu, Eileen. "China Dispatches Online Army." *ZDNet*, May 27, 2011.

Yu, Eileen. "China Dispatches Online Army: Chinese Government Confirms Existence of Cyber Warfare Unit, "Blue Army", Set Up to Improve Country's Defense Capabilities and Support Army's Internet Security Training." *ZDNET*, May 27, 2011. http://www.zdnet.com/china-dispatches-online-army-2062300502/.

Zakaria, Tabassum. "U.S. Blames China, Russia for Cyber Espionage." *Reuters*, November 3, 2011. http://www.reuters.com/article/2011/11/03/us-usa-cyber-china-idUSTRE7A23FX20111103.

Index

Lightning Source UK Ltd.
Milton Keynes UK
UKHW041254280322
400726UK00002B/8

9 781538 149669